NUCLEAR DIVISION
IN THE FUNGI

NUCLEAR DIVISION IN THE FUNGI

Edited by
I. Brent Heath

Biology Department
York University
Toronto, Ontario
Canada

1978

ACADEMIC PRESS   *New York   San Francisco   London*
A Subsidiary of Harcourt Brace Jovanovich, Publishers

ACADEMIC PRESS, INC.
111 Fifth Avenue, New York, New York 10003

United Kingdom Edition published by
ACADEMIC PRESS, INC. (LONDON) LTD.
24/28 Oval Road, London NW1 7DX

Library of Congress Cataloging in Publication Data

Mitosis Symposium, Tampa, Fla., 1977.
    Nuclear division in the fungi.

    Consists of expanded versions of papers presented in
the Mitosis Symposium at the 2d International Mycological
Congress, Tampa, Fla., Aug. 27-Oct. 3, 1977.
    Includes bibliographies and indexes.
    1. Fungi—Cytology—Congresses.   2. Mitosis—
Congresses.   I. Heath, Ian Brent.   II. International
Mycological Congress, 2d, Tampa, Fla., 1977.
III. Title.
QK601.M57   1977   589'.2'0487623   78-7437
ISBN 0-12-335950-3

PRINTED IN THE UNITED STATES OF AMERICA

CONTENTS

LIST OF CONTRIBUTORS

Arthur Forer
    Biology Department
    York University
    Toronto, Ontario, Canada

Manfred Girbardt
    Akademie der Wissenschaften der DDR
    Zentralinstitut für Mikrobiologie und Experimentelle Therapie
    Jena, German Democratic Republic

I. Brent Heath
    Biology Department
    York University
    Toronto, Ontario, Canada

Donna F. Kubai
    Department of Zoology
    Duke University
    Durham, North Carolina

PREFACE

The contributions to this book are expanded versions of papers presented in the Mitosis Symposium at the Second International Mycological Congress held in Tampa, Florida, August 27–October 3, 1977. The objectives that stimulated the organization of that symposium and led to the selection of the speakers, two of whom have not worked on fungi, were as follows: (a) to bring to mycologists a critical "state of the art" outline of the most recent information and hypotheses available from current diverse approaches to mitosis in all organisms, including the fungi, (b) to explore possible ways in which mitosis can be used as an aid to understanding fungal phylogeny, a topic much beloved by mycologists and protistologists alike. These objectives seemed to have potential appeal to a broader audience than just those able to attend the symposium. Thus, with the additional objective of bringing to nonmycologists the current information concerning fungal mitoses, the preparation of this book was undertaken.

The papers presented here do not represent an encyclopedic account of all information available on the respective topics. Rather, they are intended to highlight what the authors perceive to be the most important current information and recent major milestones. The validity of this approach, the wisdom of the authors' selection of material, and the degree of success of the authors in fulfilling the above objectives can only be judged in the light of extensive critical reading and with the benefit of hindsight some years from now.

I should like to express my gratitude to two groups of people. Dr. H. C. Aldrich and his program committee were indulgent enough to permit me to organize a mycological symposium in which only half of the speakers were mycologists. The speakers themselves deserve thanks, not only for participating in the symposium, but also for subsequently preparing their work for publication.

NUCLEAR DIVISION
IN THE FUNGI

NUCLEAR DIVISION IN THE FUNGI

HISTORICAL REVIEW
AND
INTRODUCTION

Manfred Girbardt

Akademie der Wissenschaften der DDR
Zentralinstitut für Mikrobiologie und experimentelle Therapie
Jena, Beuthenbergstr., DDR

I. INTRODUCTION

During the last 80 years fungal karyology has suffered
the same fate, though at different times, as the karyology of
protists in general. Meticulous investigators and audacious
speculants were simultaneously impressed and embarrassed by
the manyfold phenomena they were confronted with in the light
microscope. Thus it was not surprising that under the influence of evolutionary thinking at about the turn of this century attempts were made to classify the shapes of nuclei
(mainly of protozoa) into a phylogenetic succession (Hertwig,
1902, Schaudinn, 1903, Hartmann, 1911). It is to the credit
of Belar (1926) that he ended a sometimes erroneous (but
nevertheless prosperous) era by crystallizing out the facts
and separating them from pseudo-facts. It was unavoidable
that by discrediting many of the earlier results, discussion
was slowed down and a "big silence" followed in the literature. Belar (1926) reviewed his extended investigations on
the nuclei of protists and concluded that their seemingly
atypical character is dependent to a high degree on the manyfold variations of the "achromatic" structures. "... the
only characteristic of some nuclei of protists ... is the
intranuclear position of the centrosome ...". Electron microscopy has confirmed this statement and forced the thinking

again into evolutionary lines. Indeed many people are presently searching for special features of "primitiveness" in the
behaviour of the "activity centres" of the nuclei, that is the
"lokomotorische Komponente" of Hartmann (1911). (Compare his
preface of Belar's book.)

Two recent reviews (Kubai, 1975, Fuller, 1976) are available on problems dealt with in this book and the reader is
referred to them. Similar attempts have been undertaken by
Robinow and Bakerspigel (1965) and Olive (1953) so all eras
are covered with competent discussions. It is left for me to
add some historical dates from a standpoint of a cytologist
who has taken part in the discussion period which started
(again) about 20 years ago. This discussion has centered
around the question, whether fungal mitosis is only a minor
variation of a fundamental scheme or an evolutionary quali-
fied principle which makes it different from eumitosis. At
present the question is not settled and ingenious experiments
will be needed for a final resolution. Therefore, whenever
possible, future trends which should be followed (in the
opinion of this author) will be mentioned.

II. CHROMATIN

The Feulgen test for interphase nuclei was rendered more
difficult by a low DNA content and/or strong chromatin un-
coiling and hydratation. In addition many nuclei proved to be
almost completely euchromatic. Staining methods, perhaps with
lower specificity, but a more intense staining reaction had to
be found. C.F. Robinow and his coworkers developed methods
which he had applied in bacteria so masterly before and which
he now used with virtuosity for fungal nuclei. They were
based mainly on hydrolysing the specimens and either staining
them with basic dyes or mounting them in acid dyes. (Feulgen
mounted in acetocarmine; HCl-Giemsa; HCl-Orcein). Fig. 1
represents nuclei of Mucor hiemalis, it is taken from one of
the first fungal works of Robinow. The most important progress was that the huge nucleolus remained completely unstained whereas the peripheral chromatin was well stained
(Robinow, 1957).

Iron-hematoxylin-methods, well suited to and widely used
for histological work, in many cases stained only the

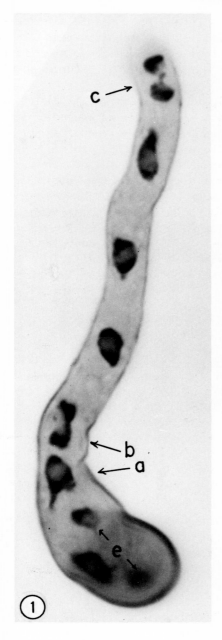

Fig. 1. Mucor hiemalis.
Germ tubes with interphase
and dividing nuclei. Chromatin is well stained in contrast to unstained nucleoli.
Fixed with acidic acid alcoh-
ol, hydrolysed with N/HCl at
60°C and stained with Giemsa
(Courtesy of Robinow, 1957
with permission of Nat. Res.
Council of Canada).

nucleolus of the fungal nucleus. The chromatin remained unstained. Early investigators, being glad to see the small
nuclei at all (Rosen, 1893, Harper, 1895, Juel, 1898) fell

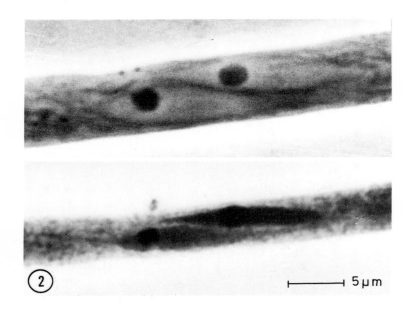

Fig. 2. <u>Trametes (= Polystictus) versicolor.</u> Interphase
nuclei in the living cell (above) as seen by phase contrast
and the same pair after fixation (CHAMPY) and staining (iron-
hematoxylin) in bright-field (below). The bright halos of the
phase-picture correspond with the chromatin (from Girbardt,
1955, with permission of VEB Fischer-Verlag, Jena).

into pardonable misinterpretation in as far as they took the
nucleolus for the whole nucleus. As vacuolized nuclei are
common in fungi, this error could be understood.

   We then showed that the iron-hematoxylin-staining depends
on the fixatives used. If the cells are fixed with fixatives
containing mercury chloride or formaldehyde the chromatin
remains completely unstained, surrounding the nucleolus with
a bright halo. But in some cases (the most dangerous ones)
the chromatin coagulates into the same granular-filamentous
structures as the cytoplasm. The nucleolus then seems to be
located immediately in the cytoplasm and chromatin cannot be
seen at all (Girbardt, 1961a). Only after application of
chrome-osmium-compounds are both chromatin and nucleolus
stained (fig. 2). A comparison of fixed nuclei with the

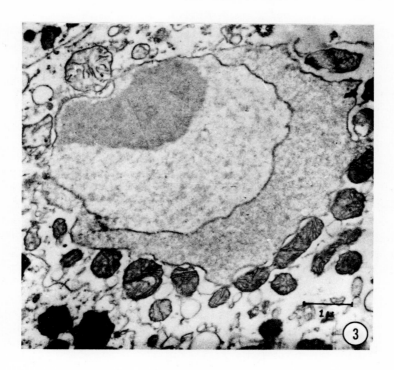

Fig. 3. <u>Allomyces</u> <u>macrogynus</u>. Section of a gamete.
Fixation: OsO₄; embedding: methacrylate. Nuclear cap, envelope, chromatin and nucleolus are distinct (Courtesy of Turian
and Kellenberger, 1956 with permission of Academic Press).

phase-contrast picture of the same nuclei in the living state
clarified what structure corresponds to chromatin and nucleolus. This was needed as in early studies of living fungal
cells similar confusions had arisen comparable to those with
the hematoxylin-staining (MacDonald, 1949, Dowding & Bakerspigel, 1954).
    So far electron-microscopy has delivered only small con-
tributions concerning specialities of chromatin within the
interphase nucleus. The reason is again the euchromatic
nature of chromatin and the difficulties in fixing fungal
cells. Areas which might correspond to heterochromatin are
very rare, though in some species (e.g. <u>Neurospora</u>, <u>Podospora</u>,
<u>Fusarium</u>, <u>Coprinus</u>) they are demonstrable (Girbardt, 1970).
    The homogenous appearance of the interphase nucleus had

been realized in the first electron micrographs (fig. 3) of
Turian & Kellenberger (1956). In this paper the standard for
good preservation with osmium derivatives was set up. Earlier
trials suffered from difficulties with permanganate (e.g.
Agar & Douglas, 1955). The wall-less gametes of Allomyces
used by Turian of course favoured good preservation, but from
this time onward electron micrographs of wall-possessing
hyphal forms also had to be measured by this quality. For
many species it was unattainable for many years.

It is striking, that near the region of the nucleus-
associated organelle (NAO = SPB) several authors have found a
heavily contrasted material possibly representing less un-
coiled chromatin (Girbardt, 1968, Heath & Greenwood, 1970,
Aist & Williams, 1972). It would be very important to apply
methods (e.g. annealing of repetitive DNA, Jones, 1970) for
identifying these parts. If "centromeric heterochromatin" in
fungi marks the position of kinetochores, this would help to
uncoil the mystery not only of this chromatin but also that of
structureless "kinetochores" found during some divisions.

The application of sensitive electronmicroscopical methods for detection of nucleic acids would also be important.
They have been developed for other organisms and allow demonstration of DNA even in viral nucleoids. They are based on
the Feulgen reaction using thallium ethylate (Moyne, 1973) or
osmium amine (Gautier & Fakan, 1974). The expanded use of
nucleases for indirect proof should also be considered. We
expect nucleic acids in the NAO (Zickler, 1973) and the
nucleolus. So far our results however are not satisfying.

Better results at the light microscopical level have been
obtained by analyzing meiosis. Chromosome morphology during
developmental steps of the generative phase was clearer much
earlier than it was in somatic divisions. This was mainly due
to the larger size of the meiotic nucleus and better stain-
ability by basic dyes. Investigations showed most improvement
after McClintock (1945) introduced the Orcein-squash-technique
in fungal karyology (Olive, 1965). Particularly, more chromosomal details have been recognized during pachytene (Singleton, 1953) than have been resolved in mitotic chromosomes.

One point of agreement exists between fungal meiosis and
mitosis: only very rarely are real metaphase plates found.
This appearance has been discussed in length for many years.
In the opinion of this author a plausible and generally
accepted explanation is wanting. All trials to interpret the

mechanism of genome separation in fungi should consider these uncertainties. These problems are discussed from another point of view by Heath in this volume.

The synaptonemal complex which connects bivalents is very important for ultrastructural research (Westergaard & v. Wettstein, 1970). Its specific structure permits unquestionable identification and helps to explain mechanisms of crossing over. It is possible also, after reconstructing all synaptonemal complexes from serial sections, to count the chromosomes (Gillies, 1972). This is important for those cases where the exact chromosome number cannot be estimated by light microscopy.

Differentiation between chromatin and nucleolar substance in the electron microscope is also important, especially in cases where the nucleolus persists during division. We have obtained good results by treating glutaraldehyde-fixed specimens with RNAase before postfixation with osmium. Chromatin is then much more contrasted by lead post-staining than are the nucleolar substances. It is the same effect as has been described for staining with basic dyes by Pollister & Leuchtenberger (1949).

III. NUCLEOLUS

It has been proven that DNA is present within the nucleolus of higher organisms, for the vertebrate nucleolus discussions lasted until 1960 (Lettré, 1956). It had been very difficult to convince opponents of the presence of such small amounts of DNA within so much RNA and protein. Again I should like to praise the accuracy of the old protistologists who had for a long time defended the existence of chromatin within the nucleolus. Though their assumption was incorrect that this was a peculiarity of only some protistean nuclei ("Karyosom", "Amphinucleolus", "Binnenkörper" etc.) it is another example of the meticulous microscopical investigations of our forefathers (Belar, 1916, Doflein, 1916).

The sharply delimited nucleolus vanishes completely in most cases after permanganate fixation, widely used even today for fixation of fungi. Its substance must be spread over the whole nucleus as the space previously occupied in the living cell cannot be detected by the electron microscope. In some cases diffuse material can be present but we have never found it in the original shape of the nucleolus.

After glutaraldehyde fixation, fibrillar and granular components embedded in a matrix can be demonstrated (Girbardt, 1970). In some species the granular component may surround the fibrillar one like shells. However it had not been possible to demonstrate nucleolus-associated chromatin. Compared with the success reached by cytochemical and biochemical analyses in describing nucleoli of higher organisms (Bernhard & Granboulan, 1968, Ghosh, 1976) the fungal nucleoli are still poorly understood. Fungi with large nucleoli, like Basidiobolus, should be suitable specimens for carrying out interesting trials. For example the mechanism of formation and extrusion of nucleolar vacuoles is still unexplained (Soudek, 1960, Erdelska, 1973).

During division nucleoli exhibit different behaviour as Pickett-Heaps (1970) has pointed out. Some "classes" have been established by him and evolutionary trends have been suggested. It would be therefore of interest to study more fungal species in this context. In Trametes (= Polystictus) the nucleolus can be largely dispersive. Its dispersion frequently occurs outside the fenestrated division in an excised part of the nucleus which is completely cut off from the chromatin-containing part. This seems to indicate that the nucleolar substance does not play a role during division. The same species however indicates that dispersion can also occur in the chromatin-containing part of the nucleus. Thus one might expect, in this case, coating of chromosomes with nucleolar material comparable to the behaviour in Spirogyra and Chara (Pickett-Heaps, 1970). So far however no morphological evidence for this assumption has been found. What should be shown by this example is that the "classes" of nucleolar behaviour must not have functional significance. Nevertheless fungal nuclear divisions would be well suited to further investigations of these questions provided better specific staining of the nucleolar substance becomes available.

Changes in nucleolar shape became evident during cinematographic recording of their behaviour in the living cell. In tissue cultures of animal cells, rotation of nuclei has been observed (Pomerat, 1953), the temporarily enclosed nucleus in the clamp of homobasidiomycetes also rotates (Girbardt, 1962a). In many cases the nucleolus was stretched in a drop-like fashion (compare fig. 4 Vph and fig. 6). This behaviour was extremely striking when the nucleus was preparing

for division. In early papers, these nuclei have been described as "Kometenkerne" when stretching had been preserved by
fixation (Ruhland, 1901, Maire, 1905, Mücke, 1908). Time-
lapse cinematography clearly indicates that the behaviour may
be explained as a response to the movement of a special point
on the periphery of the nucleus. The changing of shape therefore was interpreted as passive. It is very probable that it
is a consequence of its being connected with the activity
center, the NAO (see section V). It thus seems to be a fruit-
ful task for future cell biological work, to use the nucleolus
as an easily visible indicator for studying the behaviour of
the NAO.

IV. NUCLEAR ENVELOPE

For evolutionary and functional considerations it would
be very important to know more about fungal nuclear envelopes.
Compared with the knowledge on nuclear envelopes of higher
organisms (Franke, 1974) we are pitifully ignorant about chemical composition, pore-frequences, specific areas of chroma-
tin-attachment and enzymic activities (Tu & Malhotra, 1974).
Questions concerning the nuclear envelope and mitosis are
obvious: are fungal nuclear envelopes acting as a motor for
separation of genophores as in bacteria or dinoflagellates
(Kubai, 1975) and, if yes, how do they do this? What are the
differences between the behaviour of bacterial attachment-
points and possible chromosomal attachments with membranes?
What causes the local breakdown of the envelope in fenestrated
mitosis?
This is only a short selection of problems which might be
attacked with fungal specimens. The possibility of working
with cdc-mutants and synchronized cultures should make them
well suited.

V. NUCLEUS ASSOCIATED ORGANELLE (NAO = SPB)

A small body has been mentioned several times in the preceding parts. It seems to be a characteristic element of
fungal and other protistan nuclei. It apparently becomes
active in assembling microtubules during division, but other
functions may be fulfilled in addition. Nearly nothing is

known about its chemical composition. A few more words about this body should be delivered in an introduction, though it will play a major role in the following articles.

A. Terminology

At present, most authors use the term spindle pole body (SPB), as coined in the first International Mycological Congress (Aist & Williams, 1972). Historically there seems to be little justification to ascribe this term to Harper. In his paper (Harper, 1895), written in German and dating from his investigations under Strasburger in Bonn, he has actually coined the term but used it only once in a legend of a picture. There it refers to an artefact, probably a spindle remnant. He thought that this "body" had been drawn through the whole daughter nucleus so that it became positioned at the side of the equator. In the text he always uses "centre" for the pole-body. In a later publication he changed to "central body" (Harper, 1905).

These considerations might be of interest but of minor importance and irrelevant if they were only grumbling about a name. Then one could agree with the statement of Fuller (1976) that " ... it seems unwise to adopt a new term until we know more about the organelle's function ...". Some reasons however seem to plead for another term even at the present state of knowledge. It should fulfil the following conditions: a) It must have the ability to describe exactly all morphological variants and developmental stages by unequivocal expressions. b) It must be as neutral as possible (avoidance of provocative or prospective expressions like "kinetochore equivalent" or "centriole"). c) It must involve our knowledge that it is a cycling structure which may be governed by many factors and fulfils more than one function. d) It must indicate that it is mainly in intimate contact with the nucleus both at interphase and during division. In spite of many deliberations therefore we proposed a new term (Girbardt & Hädrich, 1975) which seems to fit well into the claims mentioned above. In the following it will be weighed against SPB.

SPB does not differentiate between developmental stages though the morphological entity can exist as a unit or part of it. In contrast NAO is defined as an organelle based on the

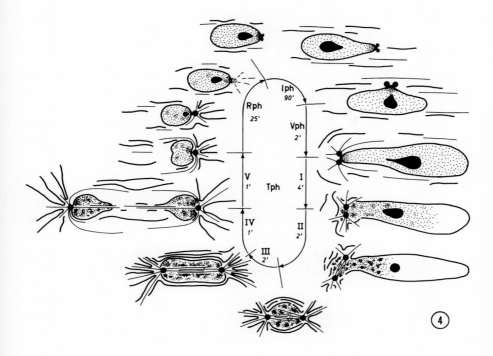

Fig. 4. Trametes (= Polystictus) versicolor. Scheme of
nuclear behaviour during the cell cycle. Iph = Interphase;
Vph = Predivision phase, corresponding to G_2 and start of prophase; Tph = Nuclear division phases (I - V, corresponding to
prophase till telophase); Rph = Replication phase of the NAO.
Shown are the components of the nucleus (chromatin, nucleolus,
envelope and NAO), microtubules assembled at the NAO, and ER-
cisternae (thicker lines), forming the "GE-field" mainly during division phases III - V. The main steps of the NAO cycle
are: Genesis during Rph; enlargement of GEs during Vph; activ-
ation for becoming a MTOC for microtubules radiating into the
cytoplasm; breakdown of the NAO-middle-plate and fenestration
of the envelope during phase I; assembling of microtubules between the two GEs during phase II; activity of GEs as motility
centres during division, phase III to V; aging of the GEs of
both daughter nuclei, loss of assembling activity for microtubules and formation of a new NAO (Unpublished).

most differentiated stage which it achieves during the cell
cycle, e.g. during interphase in basidiomycetes and many

ascomycetes (probably even in an ascomycetous yeast, Unger, 1976) it is a tripartite structure. The components of this structure are designated by special terms (globular = GE or discoid = DE entities connected by a plate-like structure, the middle part = MP). These parts are functionally and seemingly also chemically different (fig. 4). Usage of these specified terms makes it possible to accurately define what part of the whole organelle is dealt with. It is of little help to only speak of a "diglobular form" in characterizing a developmental stage. This term could correspond either to the whole organelle (with MP) or, for example during early stages of nuclear division, to two GEs without an MP (fig. 4, Tph II). SPB is in the sense of the word ("spindle") inherently bound to nuclear division. In my opinion it leads one to believe that its only function is to act as a microtubule-organizing-centre (MTOC). Moreover most investigators might expect that it would be lost during interphase. This clearly is not so in most cases. Terming it an "organelle" therefore would leave open the idea that it might also be involved in processes during interphase (e.g. in being controlled or controlling events during induction of nuclear division).

B. Early Recognition

    Again we have to start with observations made at the beginning of our century. The quality of our forefathers' observations is in many cases rather humiliating. Harper (1905) saw the "lateral granule" with almost every interphase nucleus of Phyllactinia after gentian violet staining (fig. 5). He has included it with painstaking accuracy in his water colors. Kniep (1915) saw it at the tip of a nucleus entering the clamp in Corticium. Colson (1938) even resolved the organelle as a pair of dots, again in Phyllactinia. However no comments have been possible about its function during interphase. Not so when division starts. Harper (1905) left no doubt that during prophase all "chromatinic strands" were oriented and centered with reference to the NAO. Many authors since then have described similar phenomena (e.g. Olive, 1949 for Coleosporium) but only rarely has the transparency of Harper's drawings been attained.
    A comparison with the Rabl-orientation (Rabl, 1882) in higher eukaryotes seems to be indicated but so far no

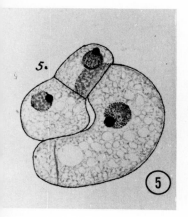

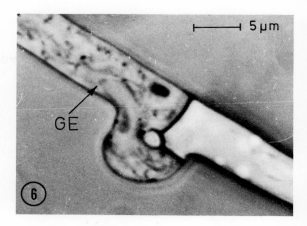

Fig. 5. <u>Phyllactinia</u> <u>corylea</u>. Interphase nuclei. In almost all the nuclei a more intensively stained granulum at the periphery has been seen and termed "central body". Staining: Gentian violet (from Harper, 1905).

Fig. 6. <u>Trametes (= Polystictus) versicolor</u>. Nucleus leaving the clamp after fusion. The globular entity (GE) precedes the movement of the whole nucleus and can be seen by phase-contrast in the living cell. The nucleolus is pulled out indicating that it is connected with the GE (from Girbardt, 1960, with permission of Springer-Verlag).

considerations on this line have come to this author's eye. The same holds for possible relations with bouquet stages (Eisen, 1900) during leptotene and pachytene. It is assumed in higher organisms, that these polarization processes originate in the cytoplasm. It might well be that also in these cases a NAO-like, not yet identified conductor on the nuclear envelope may work as well.

C. Living Behaviour

Progress was made after it became possible to demonstrate the NAO in the living fungal cell (Girbardt, 1960) but it is not possible to decide whether a complete NAO or only a GE is present. For example in fig. 6 it is now known by electron

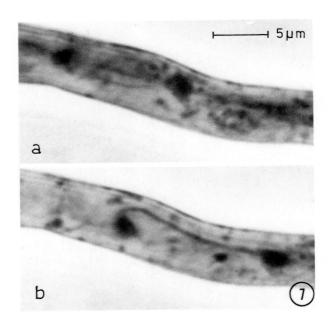

Fig. 7. <u>Trametes (= Polystictus) versicolor</u>. Two stages
of a living somatic nucleus during division. It has been
translocated within the cell by microsurgery. Nuclear envelope and central spindle can be seen (from Girbardt, 1956, with
permission of Springer-Verlag).

microscopy that the nucleus at this developmental stage (when
the nucleus is leaving the clamp) possesses only one GE. By
phase contrast-microscopy it looks like the complete NAO which
enters the clamp before nuclear division (Girbardt, 1961b).
    Another property which has become evident is the reaction
of the nucleolus to directed movements of the NAO. Fig. 6
demonstrates that the nucleolus is pulled by the prior movement of the NAO. First it was thought that a direct connection existed between the two organelles. However electron-
microscopy has proven that the nucleolus never reaches the
area of the NAO directly but ends at least about 0.5 μm behind
it. It is argued, therefore, that the behaviour of the nucleolus could be explained if it is assumed that it remains connected during interphase with its chromosomal organizer region
which in turn is positioned near the NAO to which it is

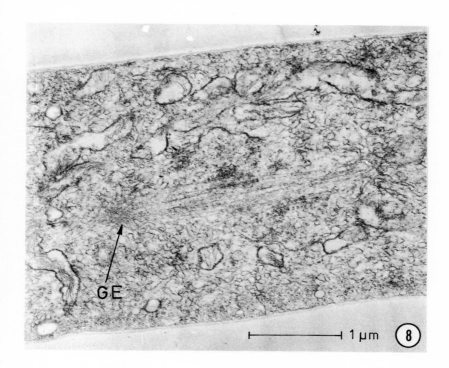

Fig. 8. <u>Trametes (= Polystictus) versicolor</u>. Dividing
somatic nucleus. Electron micrograph taken October 1962,
showing for the first time the fibrous central spindle and one
globular entity (= spindle pole body) of the nucleus-associat-
ed organelle. Fixed after Ryter & Kellenberger (Unpublished).

connected by chromatin.
    Nuclear division in the living cell was demonstrated
first after microsurgery. By this procedure the dividing
nuclei were translocated into cytoplasm with a few granules
etc. (which otherwise hide details). One of these early
attempts is shown in fig. 7. Though the nuclear envelope and
central spindle could be identified (Girbardt, 1956), the NAO
was not seen. Later, when gelatin matching the refractive index of cytoplasm (Müller, 1956) was used, both GEs at the
poles could be identified (Girbardt, 1962b, Aist, 1969).

D. Electron Microscopy

Electron microscopical results will be discussed in
length in the articles of this book. Therefore here also only
a short historical "re"-view shall be given. Technical problems have again delayed analysis. Compared with bacteria in
the early sixties, development was far behind. Isodiametric
cells (yeasts, spores) proved to be especially difficult to
fix and thin-walled, hyphal forms became so tangled that it
was practically impossible to detect nuclear division. Flat
embedding was introduced together with agar or cellophane for
keeping the hyphae in their proper positions (Girbardt, 1965).
An unpublished electron-micrograph is given here (fig. 8). It
has its own story as mentioned by Robinow and Bakerspigel
(1965, p.125). We took it in 1962 and I sent it to Dr. Robinow. To my knowledge I can agree with him that it was "the
first time that a fibrous mitotic motor device in a fungus has
been seen in the electron microscope". Dr. Robinow included
it in his lecture on nuclei of protists at the Congress for
Cell Biology, November, 1962 in San Francisco. Later it was
shown at Berkeley, Seattle and other places. So it became
public, without being published. Because we were waiting for
a better electron microscope I hoped to get still higher resolution. In addition we had just started with glutaraldehyde-
fixation. Fig. 8 was fixed following the method of Ryter and
Kellenberger (1958) but this did not preserve microtubular
elements as well as has since been possible. Therefore the
electron micrograph was not published. The same held for
pictures shown at the International Botanical Congress in
1964, one of which was published by Burnett (1968, Pl. IX).
At about the same time spindle fibers had been found in Albugo
(Berlin & Bowen, 1964) and in yeast (Moor, 1966, Robinow &
Marak, 1966). Fig. 8 indicates that the central spindle is
composed of "fibrous elements" which end in the GE. The
nuclear envelope is fragmented manyfold, surrounded by vesicu-
lar (artefactual!) ER and mitochondria. Speculations about
nucleolar origin of the "Zentralstrang" (Girbardt, 1962b) were
refuted by these results.

E. Questions of Future Interest

It is always dangerous to act as a prophet, but I suggest

that further information on the NAO will help our understanding of mitosis. Specific points which need clarification are as follows.

a) Morphological changes of the organelle during the cell cycle should be investigated with the time-dependence of these changes being explained in as much detail as possible. For biochemical work, the strains should be synchronizable. At any rate more representatives of different taxonomic groups should be studied. Comparative work might delineate more clearly the differences and similarities.

b) The biochemical constitution of the NAO should be explained. Tubulin or tubulin-associated proteins might be expected. The same holds for nucleic acids. Improvements of fractionation techniques are necessary.

c) Alterations of NAO-constituents should be expected during the cell cycle. Enzymatic activities might be present. Contributions should be possible to the question, what processes work in inducing nuclear division and to what extent is the NAO involved. Synchronized cultures are needed.

d) Attempts should be made to explain the nature of the connection between NAO, nuclear envelope and chromatin. Is the NAO a gene-product of a special chromosomal area? It might then be analogized with the nucleolus-organizer and part of the chromosome must exist outside the nucleus in species with NAO's completely extranuclear. Or is the NAO altering both membranes of the nuclear envelope in the attachment region? Both NAO and chromatin or kinetochores on the opposite side seem to stick thoroughly. Are there similarities with adhering of peripheral condensed heterochromatin in "higher" nuclei?

e) In fenestrated forms of mitosis the most exciting event is the local breakdown of the nuclear envelope. This opening might be a first evolutionary step of a process which ends in complete disintegration in higher organisms. It is important that in some cases the local breakdown of the envelope need not be coincident with assembly of microtubules. In basidiomycetes the GEs are active as MTOCs before the envelope breaks down. Ts-mutants and cytochemistry may help in answering these questions.

It is to be hoped that the original contributions collected in this book and discussed on the occasion of the symposium "Mitosis" at the Second International Mycological Congress in Tampa will help to describe the present position. It may help also to point out future trends for resolution of the question, is it worthwhile to invest further intensive work in fungal mitosis? It is obvious that at present this question remains debatable. There seems to be no doubt, if peculiarities of fungal mitosis are proven to be principally different from those of higher organisms, fighting against fungal parasites might become easier and still more effective. It seems to be evident that no single spectacular findings will explain everything but that the questions will be answered by a mosaic of results. This puzzle is far from being completed.

REFERENCES

Agar, H.D., and Douglas, H.C. (1955). J. Bact. 70, 427-434.

Aist, J.R. (1969). J. Cell Biol. 40, 120-135.

Aist, J.R., and Williams, P.H. (1972). J. Cell Biol. 55, 368-389.

Belar, K. (1916). Arch. Protistenkunde 36, 13.

Belar, K. (1926). Ergebn. Fortschr. Zool. 6, 235-654.

Berlin, J.D., and Bowen, C.C. (1964). Amer. J. Bot. 51, 650-652.

Bernhard, W., and Granboulan, N. (1968). In "The Nucleus" (A.J. Dalton and F. Haguenau, eds.), pp. 81-149. Academic Press, New York and London.

Burnett, J.H. (1968). "Fundamentals of Mycology". Edw. Arnold (Publ.) Ltd., London.

Colson, B. (1938). Ann. Bot. N.S. II, 381-402.

Doflein, F. (1916). "Lehrbuch der Protozoenkunde". Fischer-Verlag, Jena.

Dowding, E.S., and Bakerspigel, A. (1954). Canad. J. Microbiol. 1, 68-78.

Eisen, G. (1900). J. Morph. 17, 1.

Erdelská, O. (1973). Protoplasma 76, 123-127.

Franke, W.W. (1974). Int. Rev. Cytol., Suppl. 71-236.

Fuller, M.S. (1976). Int. Rev. Cytol. 45, 113-153.

Gautier, A., and Fakan, J. (1974). Experientia 30, 702.

Ghosh, S. (1976). Int. Rev. Cytol. 44, 1-28.

Gillies, C.B. (1972). Chromosoma (Berl.) 36, 119-130.

Girbardt, M. (1955). Flora 142, 540-563.

Girbardt, M. (1956). Naturwissenschaften 43, 429-430.

Girbardt, M. (1960). Planta 55, 365-380.

Girbardt, M. (1961a). Flora 150, 427-440.

Girbardt, M. (1961b). Exp. Cell Res. 23, 181-194.
Girbardt, M. (1962a). In "Handbuch der Pflanzenphysiologie",
    Bd.XIII/2, pp. 920-939. Springer.
Girbardt, M. (1962b). Planta 58, 1-21.
Girbardt, M. (1965). Mikroskopie 20, 254-264.
Girbardt, M. (1968). In "Aspects of Cell Motility" (P.L.
    Miller, ed.), Symp. Soc. Exp. Biol. 22, pp.249-259.
    Cambridge University Press, London and New York.
Girbardt, M. (1970). Z. Allg. Mikrobiol. 10, 451-468.
Girbardt, M., and Hädrich, H. (1975). Z. Allg. Mikrobiol.
    15, 157-173.
Harper, R.A. (1895). Ber. dtsch. Bot. Ges. 13, 67.
Harper, R.A. (1905). Carnegie Inst. Wash. 37, 105.
Hartmann, M. (1911). "Die Konstitution der Protistenkerne
    und ihre Bedeutung für die Zellenlehre." Fischer-Verlag,
    Jena.
Heath, I.B., and Greenwood, A.D. (1970). J. gen. Microbiol.
    62, 139-148.
Hertwig, R. (1902). Arch. Protistenkunde 1, 1-40.
Jones, K.W. (1970). Nature 225, 912-915.
Juel, H.O. (1898). Jahrb. wiss. Bot. 32, 361-388.
Kniep, H. (1915). Z. Bot. 7, 369-398.
Kubai, D.F. (1975). Int. Rev. Cytol. 43, 167-227.
Lettré, R. (1956). Zbl. Path. 95, 395-415.
Macdonald, J.A. (1949). Nature 163, 579-580.
Maire, R. (1905). C.R. Soc. Biol. Paris 58, 726-728.
McClintock, B. (1945). Amer. J. Bot. 32, 671-678.
Moor, H. (1966). J. Cell Biol. 29, 153-155.
Moyne, G. (1973). J. Ultrastr. Res. 45, 102-123.
Mücke, M. (1908). Ber. dtsch. Bot. Ges. 26a, 367-378.
Müller, R. (1956). Mikroskopie 11, 36-46.
Olive, L.S. (1949). Amer. J. Bot. 36, 41-54.
Olive, L.S. (1953). Bot. Rev. 19, 439-586.
Olive, L.S. (1965). In "The Fungi" (G.C. Ainsworth and A.S.
    Sussman, eds.), Vol. 1, pp. 143-161. Academic Press,
    New York and London.
Pickett-Heaps, J.D. (1970). Cytobios 6, 69-78.
Pollister, A.W., and Leuchtenberger, C. (1949). Nature 163,
    360-361.
Pomerat, C.M. (1953). Exp. Cell Res. 5, 191-196.
Rabl, C. (1882). Morph. Jahrb. 10, 214-298.
Robinow, C.F. (1957). Canad. J. Microbiol. 3, 771-789.
Robinow, C.F., and Bakerspigel, A. (1965). In "The Fungi"

(G.C. Ainsworth and A.S. Sussman, eds.), Vol. 1, pp. 119-142. Academic Press, New York and London.

Robinow, C.F., and Marak, J. (1966). J. Cell Biol. 29, 129-151.

Rosen, F. (1893). Cohn Beitr. Biol. Pflanzen 6, 237-266.

Ruhland, W. (1901). Bot. Zeitg. 59, 187-206.

Ryter, A., and Kellenberger, E. (1958). Z. Naturforsch. 13b, 597-605.

Schaudinn, F. (1903). Arb. Reichsgesundheitsamtes Berlin 19, 547-576.

Singleton, J.R. (1953). Amer. J. Bot. 40, 124-144.

Soudek, D. (1960). Exp. Cell Res. 20, 447-452.

Tu, J.C., and Malhotra, S.K. (1974). J. Histochem. Cytochem. 21, 1041-1046.

Turian, G., and Kellenberger, E. (1956). Exp. Cell Res. 11, 417-422.

Westergaard, M., and vonWettstein, D. (1970). C.R.Lab.Carlsberg 37, 239-268.

Zickler, D. (1973). Histochemie 34, 227-238.

CHROMOSOME MOVEMENTS DURING CELL-DIVISION:
POSSIBLE INVOLVEMENT OF ACTIN FILAMENTS

Arthur Forer

Biology Department
York University
Downsview, Ontario

I. INTRODUCTION

In this chapter I concentrate on the question of whether or not actin is involved in producing force for chromosome movements during cell-division. The data to date concern whether or not actin is a genuine spindle component: these data will be considered in detail. By way of introduction, I first review some of the basic and generally agreed upon aspects of mitosis, such as the forces involved, the role of chromosomal spindle fibres, and so on. Then I briefly summarize the main hypotheses in which microtubules are considered to be the force producing agents, and I briefly summarize my view of the status of these hypotheses. Finally, I discuss critically and in detail the electron microscopic and light microscopic data which suggest that actin is a component of chromosomal spindle fibres. I discuss also the negative evidences, and I discuss the counter-arguments that are used to argue that actin is not a component of spindles. I conclude that actin is probably a spindle component, and is probably involved in chromosome movement. However, none of the evidences for actin being a spindle component are unequivocal, and I discuss the kinds of data that need to be obtained to substantiate that actin is a spindle fibre component and that actin is involved in producing the force for chromosome movement.

21

II. BASIC POINTS ABOUT MITOSIS

The basic points reviewed here are discussed in various reviews (Schrader, 1953; Mazia, 1961; Inoué, 1964; Inoué & Sato, 1967; Forer, 1969, 1974; Nicklas, 1971, 1975; Inoué & Ritter, 1975; McIntosh et al., 1975), to which readers are referred for detailed discussion.

In anaphase, chromosomes move from the equator towards the poles; this is accomplished either by shortening of the chromosome to pole distance, as the poles remain a constant distance apart, or by separation of the poles as the chromosome to pole distance remains constant, or by simultaneous shortening of chromosome to pole distances and lengthening of pole to pole distances. Different situations apply in different cells, or at different times in the same cells (e.g. Ris, 1943, 1949; Jacquez & Biesele, 1954; Forer, 1966; Fuseler, 1975). These motions may be more than just descriptively different, for, as discussed by Ris (1943, 1949) and Jacquez & Biesele (1954), and as emphasized by McIntosh et al. (1975) and by McDonald et al. (1977), for example, the different motions may have different mechanisms (e.g., chromosome-to-pole motion might be caused by microtubule cross-bridges interacting with microtubules to cause sliding, whilst the poles might move apart because of force produced by actin filaments, as mentioned by McIntosh et al., 1975; or the converse might apply). The discussion in this chapter deals exclusively with movements which occur by shortening of the chromosome-to-pole distances.

Chromosomes move polewards with velocities in the range of 0.2 - 5.0 μm/min., much slower than protoplasmic movements, which are in the range of μm/sec. (see Wolpert, 1965; Nicklas, 1975); chromosome movements are at very slow velocities indeed, for they are very close to the velocities of tectonic plate movements! The forces for the movements are also very small: 1 actin filament together with 1 myosin filament, producing force at the efficiency of actin-myosin filament interactions in skeletal muscle, produce 3,000 times more force than necessary in order to move a chromosome to its pole (see Forer, 1969; Nicklas, 1971, 1975; Gruzdev, 1972). Likewise, one ATPase molecule (e.g., one arm on a ciliary doublet) working for one second produces more than enough energy to move a chromosome polewards (see Forer, 1969; Nicklas, 1975). Taking considerations like these into account, several authors have suggested that there is a rate limiting step, a 'governor', separate from the force production step, and that microtubule depolymerisation might be that rate limiting step: the slow depolymerisation of microtubules limits the rate of chromosome movement, which, without this impedance, would be much faster (e.g., McIntosh et al., 1969; Nicklas, 1971, 1975; Forer, 1974).

As a chromosome moves poleward the chromosome's kinetochore (spindle-fibre-attachment region) generally leads the way. As seen in fixed and stained preparations or as seen in the polarising microscope, some spindle fibres extend from kinetochores to poles while other spindle fibres do not end at kinetochores: these two kinds of light-microscopically visible spindle fibres are called chromosomal spindle fibres and continuous spindle fibres, respectively, using terminology of Schrader (1953). As a chromosome moves polewards the associated chromosomal spindle fibre shortens without measurable change in width.

The force to move a chromosome appears to be applied at the kinetochore, is directed towards the pole, and is transmitted to the chromosome by the chromosomal spindle fibre (Cornman, 1944; Mazia, 1961; Nicklas, 1971). The evidences for this are in part negative, in that all other hypotheses seem to be ruled out (Cornman, 1944; Taylor, 1965; Gruzdev, 1972), and are in part positive, in that there is indeed a mechanical link between chromosome and pole (Nicklas & Staehly, 1967) and in that ultraviolet microbeam irradiation of single chromosomal spindle fibres can stop chromosome movement whilst equivalent doses applied elsewhere in the cell do not stop chromosome movement (Forer, 1966). Hence chromosomal spindle fibres seem to be necessary for chromosome movements in anaphase; there is not enough data to say whether the chromosomal spindle fibres are both necessary and sufficient, but most workers agree that the forces for chromosome movement are transmitted to the chromosome by the chromosomal spindle fibre.

Spindle fibres are birefringent. Birefringent materials by definition have indices of refraction which are different for light polarised in different directions, and it is this optical property of spindle fibres which allows them to be seen and studied in living cells (Swann, 1951 a, b; Inoué, 1953, 1964). It should be emphasized that spindle fibres contain components which do not contribute to the birefringence,

but which could contribute to producing the force: spindle
fibre birefringence does not arise from all the components in
a light microscopically observed spindle fibre. Hence, when
one says that the force for chromosome movement arises from
(or is transmitted by) the chromosomal spindle fibre, this
does not necessarily mean that the force for chromosome movement arises from (or is transmitted by) the birefringent component of the chromosomal spindle fibre (see, e.g., Forer,
1976). To restate this somewhat, the birefringence of the
spindle arises from material which occupies only 2 - 3% of the
volume of a spindle fibre (Sato et al., 1975; Forer, 1976);
the question is, whether the force comes from this birefringent material or rather from some other component in the remaining 97 - 98% of the volume. There are data which support
the hypothesis that the force arises from a component which
contributes little to the birefringence (Forer, 1966, 1969),
but Inoué and co-workers (e.g., Inoué & Ritter, 1975), on the
other hand, assume that the birefringent component gives rise
to the force. Be that as it may, such considerations auto-
matically raise the issue of what exactly a spindle fibre is
composed of, which I now discuss.

It is important to realize that the chemical composition
of the mitotic spindle and of individual spindle fibres is not
known, even though the authors of many text books and many research articles seem to use 'spindle microtubule' as a synonym
for 'spindle fibre'. The reason that we do not know the chemical composition of the spindle is that about 90% of the dry
matter of the spindle is lost during the isolation of spindles
from cells (Forer, 1969; Forer & Goldman, 1972), as summarized
in Table I. How much of the lost material could be microtubule protein (tubulin)? One can deduce that microtubules and
tubulin, combined, make up only a small fraction of the dry
matter in a spindle, as follows. In analyzing the isolated
mitotic apparatus (MA, the chromosome-spindle-aster complex),
it is found that microtubules make up only 10% of the mass
(Cohen & Rebhun, 1970; Bibring & Baxandall, 1971). Since the
isolated MA contains only 10% of the dry matter of MA in vivo,
this means that microtubules are only about 1% of the MA mass,
that 9% (the remainder of the 10%) is different from microtubules, and that about 90% of the MA mass is unknown. Could
this 90% be soluble tubulin which is lost during the isolation? If so, one could say that spindles are composed primarily (90%) of soluble tubulin in equilibrium with the poly-

TABLE I

Losses During Isolation of Mitotic Apparatus (MA)
from Sea Urchin Zygotes

MA IN VIVO (INTERFERENCE MICROSCOPY)

Material Studied	Dry Matter (gm/100 cm.^{3})	Reference
Zygotes of Psammech- inus miliaris. This is measurement of asters.	21	Calculated from data of Mitchison & Swann (1953), as described in Forer & Goldman (1972) (see Table 6).
Zygotes of Psammech- inus miliaris		
asters	21	Forer & Goldman (1972)
spindle	20	
Zygotes of Echinus esculentus		
asters	19	Forer & Goldman (1972)
spindle	18	

ISOLATED MA

Material Studied	Dry Matter (gm/100 cm.^{3})	Reference
Various species, using various methods of isolation	0.75 - 1.5	Determined chemically by various authors, summarized in Forer (1969) (see Table 8).
From Psammechinus miliaris and Echinus esculentus, isolated using 1M hexylene glycol	2-6 (depen- ding on the pH of the isolation)	Determined by inter- ference microscopy (Forer & Goldman, 1972)

merised form, microtubules. The answer to this question, however, is 'no'. One argument which demonstrates this is based
on birefringence measurements: one assumes that spindle birefringence (retardation) is a measure of the amount of polymerised tubulin (microtubules) relative to non-polymerised tubulin (Inoué & Ritter, 1975; Inoué et al., 1975), one assumes
that the minimum amount of tubulin is polymerised into microtubules, one assumes that the microtubules are preserved during the isolation but that the soluble tubulin is lost during
isolation of the MA, and one then calculates the amount of
tubulin lost. One concludes from such an exercise that tubulin is at most 25% of the dry matter of the MA. [To present
this in detail, Inoué and co-workers suggest that spindle birefringence (retardation) is a measure of the amount of microtubules relative to non-polymerised tubulin, and that when birefringence (retardation) is measured at different temperatures the maximum birefringence is attained when all the tubulin is polymerised into microtubules (e.g., Inoué, 1964; Inoué
& Sato, 1967; Inoué & Ritter, 1975; Inoue et al., 1975). Data
on MA birefringence relative to the maximum birefringence
attainable can then be directly converted into relative amounts of microtubules vs non-polymerised tubulin. Accepting
their hypotheses, for the purpose of this discussion, relevant
data are given by Stephens (1973), who studied birefringence
(retardation) of MA in vivo in zygotes of the sea urchin,
Strongylocentrotus droebachiensus. When zygotes were grown at
various temperatures, Stephens found a 50% increase in spindle
birefringence (retardation) when the temperature was raised
above the temperature that was optimum for growth; the birefringence at the lower limit of growth was 10% of the maximum
birefringence. This means, then, that at the lower limit of
growth there is 10 times more soluble tubulin than microtubules. In those studies in which microtubules were estimated
as 10% of the mass of isolated MA (Cohen & Rebhun, 1970, for
MA from zygotes of the sea urchin Arbacia punctulata; and
Bibring & Baxandall, 1971, for MA from zygotes of the sea urchin Strongylocentrotus purpuratus), the zygotes were grown
near the optimum temperature. But even if one were to argue
that the zygotes were grown at the lower limit of temperature,
then, assuming that there would be 10 times more monomer (tubulin) than polymer (microtubules) in the MA in vivo, and that
all the monomer (tubulin) was lost during isolation, since the
polymer (microtubules) is only 1% of the in vivo dry matter

(i.e., 10% of the 10% of the dry matter left in the isolated MA), the monomer plus polymer together comprise less than 20% of the in vivo dry matter (i.e., 11 times 1%).] Another argument which demonstrates that tubulin is less than 25% of the MA dry matter is based on measurements of the total amount of tubulin per cell: one uses chemical measurements of total tubulin per cell (or of fraction of the cell protein which is tubulin), interference microscope measurements of relative concentrations of dry matter in spindle vs cytoplasm, and cytological measurements of relative volumes of spindle vs cytoplasm; then, by assuming that all the cell's tubulin is in the volume occupied by the spindle, one calculates the maximum amount of the spindle dry matter which could be tubulin, which, too, is less than 25% (A. Forer, in preparation). Thus, tubulin is a minority fraction of the dry matter in a spindle, and one just does not know the chemical composition of $\geq 65\%$ of the dry matter in spindles.

To summarize, spindle fibres produce (or, at least, transmit) force for chromosome movement. Spindle fibres are birefringent, but the birefringence arises from a small volume fraction of the spindle and, in my view, it is not known whether the birefringent fraction or a different fraction gives rise to the force. The chemical composition of the spindle is unknown, because most of the material in in vivo spindles is lost during isolation. Microtubules and tubulin, combined, make up a minority fraction of the spindle, and the composition of the remainder is largely unknown.

One last point to be considered is the source of spindle birefringence. I have argued previously (Forer, 1976) that chromosomal spindle fibre birefringence is most likely due to the birefringence of microtubules plus birefringence due to other components, and that spindle birefringence in other parts of the spindle is due to the birefringence of microtubules minus birefringence due to yet other components. Some other workers disagree, however; Inoué (1976), for example, considers that the "problem is now pretty much resolved", and that, without doubt, spindle birefringence is due solely to microtubules. Inoué & Ritter (1975) reach similar conclusions. I will discuss the evidences on this point in detail elsewhere (Forer, Int. Rev. Cytology, in preparation). I here reiterate that there are strong evidences suggesting that several components contribute to spindle birefringence (e.g. Forer et al., 1976) and that the evidences adduced by Inoué

(1976) and Inoué & Ritter (1975) are equivocal. As one example of this, I consider the data of Sato et al. (1975) which
are interpreted by them as proof that the birefringence of the
isolated mitotic apparatus (MA) is due solely to microtubules.
These authors isolated MA from oocytes of the sea star Pisast-
er, using hexylene glycol, and they measured the form birefringence of the MA by imbibing the MA with media of different
refractive indices. From the measured form birefringence
curve they calculated the volume fraction of rods which theor-
etically would account for the birefringence, and this was
about 2%. Electron microscopic observations of isolated MA
showed that microtubules occupied 2% of the volume, exactly
the same fraction of the volume as predicted by the form birefringence measurement. They concluded that the birefringence
of the isolated MA is due solely to microtubules. For several
reasons, however, this argument is not conclusive. Consider,
for example, the following argument concerning their experimental protocol. Sato et al. (1975) measured birefringences
of MA that were fixed with glutaraldehyde, were dehydrated
through a series of dimethylsulphoxide-hexylene glycol solutions, and then were imbibed with various organic solutions,
each of which had a different refractive index. For electron
microscopic observations, however, the MA were treated differently from those in which birefringence measurements were
made: they made electron microscopic observations on MA which
were fixed in glutaraldehyde, post-fixed in osmium-tetroxide,
dehydrated through a series of ethanol solutions, and then embedded and sectioned. The two protocols, summarized in Table
II, do indeed differ; but are the differences significant?
Let us consider the one difference, the presence of osmium
tetroxide in the electron microscopic protocol and its absence
in the polarisation microscopic protocol. Forer & Blecher
(1975) studied crane fly spermatids and found that microtubules are seen clearly when glutaraldehyde fixation is followed by osmium tetroxide post-fixation, but that microtubules
are no longer seen when the osmium-tetroxide post-fixation is
omitted. There were no electron microscopic data in Sato et
al. (1975) on MA treated without osmium tetroxide post-fixation, so it is possible that in fact no microtubules as such
were even present in the MA in which they measured form birefringence. To see if this might be true, I isolated MA from
sea urchin zygotes, using hexylene glycol, and fixed the MA
with glutaraldehyde; without osmium tetroxide post-fixation I

TABLE II

Difference Between Polarisation Microscopy and
Electron Microscopy Protocols

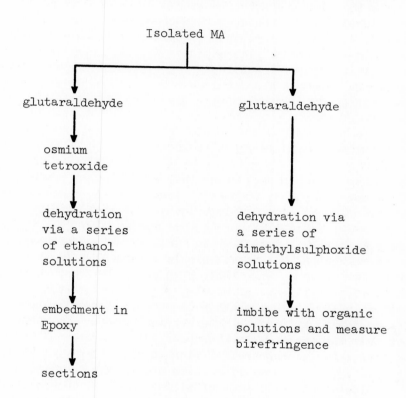

Isolated MA

glutaraldehyde

osmium
tetroxide

dehydration
via a series
of ethanol
solutions

embedment in
Epoxy

sections

electron
microscopy

glutaraldehyde

dehydration via
a series of
dimethylsulphoxide
solutions

imbibe with organic
solutions and measure
birefringence

polarisation
microscopy

dehydrated the MA via a series of dimethylsulphoxide-hexylene glycol solutions (using the protocol of Sato et al., 1975), embedded the MA in Araldite, and studied serial sections of several MA, in the electron microscope (Forer, in preparation): I saw no microtubules. Hence, even on these grounds alone, the arguments of Sato et al. (1975) concerning isolated MA are equivocal.

Regardless of the exact nature of the microtubule contribution to spindle fibre birefringence, in most eukaryotes microtubules are associated with chromosomes, at their kinetochores, and hence microtubules would seem to have some role in chromosome movement. The question is, what exactly is the role of spindle microtubules? I now consider several hypotheses in which it is assumed that microtubules produce (and transmit) force for chromosome movement.

III. MICROTUBULES AS FORCE PRODUCERS

Several hypotheses state that microtubules produce the force for chromosome movement. Some authors consider that forces arise by slow depolymerisation of microtubules (Dietz, 1969, 1972; Inoué & Ritter, 1975; Gruzdev, 1972). Other authors consider that forces arise by interactions between parallel microtubules, via cross-bridges extending between the tubules (McIntosh et al., 1969; Nicklas, 1971), and that, as in the sliding filaments of muscle, forces are exerted by microtubules which slide against each other. Yet others consider the forces to arise from interactions between microtubules, but that the forces arise by non-sliding ("zipping") interactions between non-parallel microtubules (Bajer, 1973; Bajer et al., 1975). I summarize these hypotheses in turn.

Slow depolymerisation of microtubules is considered by Dietz (1969, 1972) to be the motive force for chromosome movement: microtubules extend between chromosomes and poles, anchored at both ends; some microtubule subunits (tubulin) leave each microtubule at various (random) places along the entire length of the microtubule, and immediately the remainder of the microtubule subunits rearrange and fill in the gaps. This results in shortening of the microtubules, at constant microtubule width, and this shortening of the microtubules by slow depolymerisation and rearrangement provides the force to move the chromosomes. Inoué (1964) and Inoué & Ritter (1975) also

suggest that slow depolymerisation of microtubules provides
the force for chromosome movements and that the subunits which
remain in the microtubule rearrange (move) and fill in the
gaps after subunits leave the microtubule; the difference is
that in the Inoué & Ritter (1975) hypothesis this slow depolymerisation is considered to take place exclusively at the
poles or near the poles, rather than along the entire microtubule length as Dietz (1969, 1972) postulates. Gruzdev
(1972) suggests a similar hypothesis, but in this case the
slow depolymerisation takes place primarily at (or near) the
kinetochores.

      Kinetochore-associated microtubules undoubtedly shorten
as chromosomes move polewards, as in the previously described
hypotheses, but the question is whether this shortening
causes chromosome movement, or whether something else causes
movement and shortening is a consequence of movement. In
hypotheses of McIntosh et al. (1969) and of Nicklas (1971),
bridges between parallel microtubules exert the forces which
cause chromosomes to move, and microtubule shortening is a
consequence and not the cause of movement. In the hypothesis
of McIntosh et al.(1969) some of the spindle microtubules
arise (grow) from spindle poles and others arise (grow) from
kinetochores. The former become interpolar ('continuous')
microtubules, and do not attach to chromosomes, whilst the
latter become kinetochore microtubules and extend from kinetochore to pole. According to the hypothesis, each microtubule
has an inherent polarity which is a function of the direction
of growth of the microtubule, and this polarity causes force
to be exerted towards the origin of growth of each microtubule (either kinetochore or pole). The force is exerted by
microtubule-associated cross-bridges, and net force, which
produces sliding, arises only when cross-bridges on any given
microtubule interact with cross-bridges on microtubules with
opposite polarity. Thus, if one considers a kinetochore facing a pole, the microtubules from that kinetochore slide polewards (and pull the attached chromosome polewards) by interacting with microtubules which arose (grew) from that pole; no
net sliding force is exerted if the kinetochore microtubules
interact with microtubules which arose from the opposite pole.
According to the hypothesis shortening of kinetochore-to-pole
microtubules is due to disassembly of microtubules at the
pole, and, again by hypothesis, the kinetics of this microtubule disassembly in part defines the rate of chromosome

movement. Similar arguments explain spindle elongation:
microtubules from one pole will interact with microtubules
from the other pole to produce force (sliding) which pushes
the poles apart; but microtubules from the same pole will not
interact with each other to produce force (nor will microtubules from the same kinetochore). Nicklas (1971) critically
evaluated this hypothesis and modified it to better account
for various experiments which did not fit the model. Nicklas
(1971) proposed that only microtubules arising from poles have
associated cross-bridges, that is to say, that microtubules
arising from kinetochores do not have cross-bridges. In the
Nicklas model the cross-bridges produce force towards the pole
to which they are attached, but the sliding forces which they
exert are exerted only when activated by kinetochore microtubules of opposite polarity; the resulting forces cause sliding
between microtubules such that a chromosome is moved towards
the pole to which its kinetochore faces, but no force is exerted towards the opposite pole.

In the final hypothesis to be considered, non-sliding
("zipping") interactions between non-parallel microtubules are
assumed to provide the motive force for chromosome movement,
in the "zipper hypothesis" (Bajer, 1973, and Bajer et al.,
1975). "Zipping", defined by Bajer (1973), occurs between
microtubules which are initially at an angle to each other
(e.g., 30°): the end of one ('movable') microtubule initially
bonds to another ('fixed') microtubule. The two microtubules
then bond laterally, in increasing amounts, and eventually become parallel and tightly bound. The lateral bonding occurs
"progressively along their length..." and as a result the non-
bonded "free end..." of the 'movable' microtubule moves laterally towards (i.e. roughly perpendicular to) the 'fixed'
microtubule. Thus, "zipping" is basically a lateral movement
of one microtubule due to lateral interactions between it and
an initially non-parallel microtubule. According to the hypothesis, such interactions can cause the free end of the 'movable' microtubule to move longitudinally (i.e., in the direction of the length of the 'fixed' microtubule) if there are constraints which prevent the region near the 'free end' of the
'movable' microtubule from moving laterally. In such a case
the 'bound end' of the 'movable' microtubule would bond laterally; then the bonding progresses towards the 'free end' and
in regions where no lateral movements can occur this constraint causes the 'movable' microtubule to bend, and the

bending, in turn, causes the 'free end' to move longitudinally. By attaching the chromosome's kinetochore to the 'free
end', one thereby, by hypothesis, has "zipping" move the
chromosome polewards. A constraint to lateral movement which
would force longitudinal movement of "free" (kinetochore associated) ends of a microtubule could occur, according to Bajer
(1973), if interpolar microtubules are adjacent to the 'free
end' to prevent it, physically, from moving laterally; another
constraint is if two microtubules which are attached to the
same kinetochore 'zip' in opposite lateral directions. Lateral movement is prevented with either constraint and, according
to the hypothesis, microtubules bend in the region of the lateral constraint, which in turn results in the kinetochore (and
hence the chromosome) being transported polewards. Stated in
another way, the hypothesis says, for example, that if all
non-kinetochore microtubules are parallel, in a pole-to-pole
direction, then kinetochore microtubules which are at an angle
to the non-kinetochore microtubules interact with the non-
kinetochore microtubules to "zip", and lateral restraints con-
vert "zipping" into polewards movement. Complete movement of
a chromosome to a pole (maximum shortening of kinetochore
microtubules) is a result of a series of such "zippings",
amounting to "hundreds, if not thousands..." (Bajer et al.,
1975). In common with previous hypotheses (e.g. McIntosh et
al., 1969; or Forer, 1974), Bajer et al. (1975) assume that
"assembly-disassembly of MTs [microtubules] is a factor regul-
ating ... the rate of movements..."

Several hypotheses for chromosome movement have been summarized in this section. Different hypotheses assume different roles for spindle microtubules, though in all hypotheses
spindle microtubules (with or without associated cross-
bridges) produce force for chromosome movement. How is one to
decide which, if any, of these hypotheses is correct? One can
look at the experimental bases of each hypothesis, or one
can look at the phenomena explained by the hypothesis and
the phenomena which are difficult to explain by each individual hypothesis. This is standard procedure, and I will elsewhere review these hypotheses in this manner (A. Forer, Int.
Rev. Cytol., in preparation). For the purpose of this discussion, I assert that, in general, the hypotheses are very
broad and are used to explain many facets of mitosis; furthermore, data which argue against some aspects of individual
hypotheses do not necessarily rule out other aspects of the

same hypotheses. For example, there seem to be no interactions possible between kinetochore microtubules and interpolar microtubules in some spindles (e.g., Heath, 1974;
McDonald et al., 1977); while this would seem to argue against
zipping or cross-bridges causing chromosome-to-pole movements,
the data do not rule out the idea that cross-bridge action
causes pole-to-pole elongation. The difficulty is that the
data do not allow one to rule out or rule in any of the hypotheses; nor, indeed, do any data demonstrate unequivocally
that microtubules produce the force, or even transmit the
force to the chromosomes. This is emphasized by the points,
summarized above, illustrating that we do not really know
what spindle fibres are composed of, or what the birefringence
is due to. One would guess that until we get these data, or,
more importantly, that until we can chemically dissect an in
vitro model system or until we can genetically or otherwise
rigorously dissect in vivo spindles, that we will not be able
to accept or reject any of the above hypotheses.

I now discuss the hypothesis that actin produces the
force for chromosome movement. In my view, the poleward movement of chromosomes is more likely to be due to processes involving actin than to microtubules functioning as predicted by
any of the microtubule hypotheses summarized above.

IV. ACTIN AS FORCE PRODUCER

A. General Role of Actin

Actin is found in cells throughout the plant and animal
kingdoms, in both muscle and non-muscle tissue (reviews in
Pollard & Weihing, 1974; Forer, 1974, 1978; Hepler & Palevitz,
1974), and is even found in some prokaryotes (Niemark, 1977;
Wang et al., 1976). A generalization which follows from these
data is that actin is involved in all cellular motile systems,
as argued elsewhere (e.g., Behnke et al., 1971; Forer &
Behnke, 1972a, 1972b; Forer, 1974, 1978; Pollard & Weihing,
1974; Pollard, 1977). This is similar to the generalizations
that ATP is the universal energy source, and that DNA is the
universal genetic system: both of these statements have exceptions, but they remain the general statements that one
accepts for any new system until there are data which prove

otherwise. I believe that the same can be said for actin and
cell motility; that is, that for any given motile system one
expects actin to be involved, unless and until there are data
which demonstrate otherwise, such as in cilia and flagella
(Gibbons et al., 1976) or vorticellid spasmonemes (Routledge
et al., 1976).

In overview, an argument for this generalized role of
actin would run as follows. Actin-myosin interactions produce
force causing muscle contraction (e.g., Huxley, 1972). Actin-
myosin interactions most likely produce force causing amoeboid
motions (e.g., Taylor et al., 1973; reviews in Forer, 1974;
Pollard & Weihing, 1974; Allen & Taylor, 1975; Taylor, 1976).
Actin filaments are involved in protoplasmic streaming in
algae (e.g., Palevitz, 1976); in algae the polarity of the
actin filaments is directly correlated with the direction of
streaming (Kersey et al., 1976). Cell cleavage seems to occur
by means of actin filaments in the contractile ring, in the
cleavage furrow (Perry et al., 1971; Forer & Behnke, 1972c;
Schroeder, 1973, 1975, 1976; Gawadi, 1974), perhaps in consort
with myosin filaments (Fujiwara & Pollard, 1976). Actin is
involved in acrosome movements (review, Tilney, 1975), in
movements of microvilli (Mooseker, 1976), and in movement of
animal cells (e.g., Goldman et al., 1976). Actin filaments
are found in eukaryote cells ranging from man to amoebae to
algae to higher plants (review, Pollard & Weihing, 1974;
Forer, 1974, 1978; Hepler & Palevitz, 1974), in places where
one expects force production. What more reasonable generalization than to assume that this universal and highly conserved protein is involved in cell motility in all these systems?
This is not to say that one knows how actin is involved in any
one system (except, perhaps, skeletal muscle), or which systems for sure are exceptions, or what other roles actin might
have (e.g., Lazarides & Lindberg, 1974), but rather that one
makes this generalization and then uses the large body of data
on muscle proteins as a guide to studying these other systems
(e.g., Huxley, 1976). What about chromosome movements?

From the 'generalization' just described, and in view of
the uncertainties reviewed in Section II, about spindle composition and about which spindle components produce force, one
assumes that actin is involved in producing force for chromosome movements. Before I summarize and discuss data which
deal with this point I want to state explicitly that the data
to date deal primarily with whether or not actin is present in

spindles. If actin is **absent** from spindles, one can rule out
the possibility that actin is involved in chromosome movement.
The converse, that actin is **present** in spindles, only is consistent with the possibility, and does not prove it: only
further work can do that. But if actin **is** present in spindles
then one argues from this, and from the generalized role of
actin, that actin is probably involved in producing force for
chromosome movement. I now discuss the detection of actin in
the spindle as determined using various methods, namely, electron microscopy, fluorescently labelled antibodies, and fluorescently labelled HMM, discussing each method in turn.

B. Electron Microscopic Studies of Actin in the Spindle

     I first discuss the general problem of detecting actin
electron microscopically. Then I discuss the specific data
which deal with actin in spindles.

1. Electron Microscopic Detection of Actin

     Actin exists as monomeric, globular actin (G-actin), as a
non-globular but bound actin (e.g., Tilney, 1975), or as filaments (F-actin) that often contain other components, such as
tropomyosin and troponin. Actin-containing filaments from
muscle are 6-8 nm in diameter as seen in both negatively
stained preparations and in sections, and can be identified as
containing actin by means of their reaction with heavy meromyosin or with heavy meromyosin subfragment 1 (review in Pollard
& Weihing, 1974; Forer, 1978): actin-containing filaments <u>from</u>
<u>muscle</u> have clearly delineated edges and are 6-8 nm wide (Fig.
1), whereas actin-containing filaments that have reacted with
heavy meromyosin (HMM) or heavy meromyosin subfragment 1 (S1)
have a series of arrowheads down the length of each filament,
or at least appear 'decorated', with not clearly delineated
edges, and are more than 20 nm wide (Fig. 2). The arrows all
point in the same direction on any given filament. Actin-containing filaments <u>from non-muscle cells</u> appear identical to
those from muscle cells, and react with HMM (or S1) from rabbit skeletal muscle to form a series of arrowheads, all of
which point in the same direction on any given filament. The
arrowhead complexes are identical to those formed from rabbit
skeletal muscle actin, and are found when rabbit skeletal

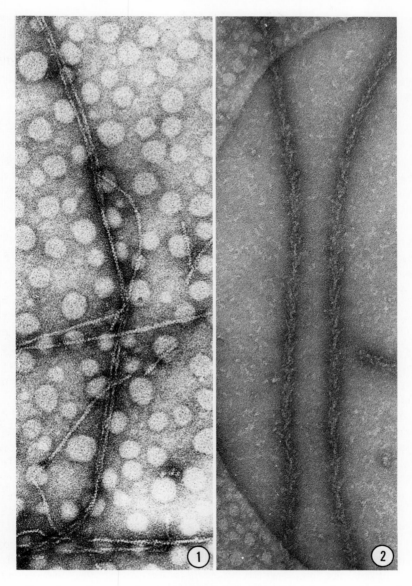

Fig. 1. An electron micrograph of negatively stained rabbit skeletal muscle actin. Fig. 2 is an electron micrograph of negatively stained rabbit skeletal muscle actin that was reacted with S1, to give arrowheads. (Techniques are given in Forer, 1978). In Fig. 2 the arrows all point downwards. Both are at the same magnification (X 190,000).

muscle HMM is mixed with actin from a broad range of non-
muscle cells, from man to amoebae to algae to higher plants.

The 'arrowhead' appearance with HMM or S1 is specific for
actin-containing filaments, as shown by two kinds of evidence.
First, a 'theoretical' argument, namely that the arrowhead
appearance is due to the arrangement of the G-actin monomers
in the filament: the F-actin filament contains 2 chains of
G-actin monomers that twist around each other, and it is this
arrangement of monomers that gives rise to the arrowheads
(Moore et al., 1970), as follows. There is one HMM binding
site per actin monomer, and the binding sites on different
monomers twist around the long axis of the filament as the two
actin chains twist around. The arrowhead appearance is due to
the superposition of the images of HMM (or S1) molecules
which are at different angles (with respect to the filament
long axis) at different positions along the length of the filament (see Moore et al., 1970, and Huxley, 1973). Thus, since
no other known filament has both a similar arrangement of
monomers into 2 chains and one HMM binding site per monomer,
no other filament will give arrowheads. Second, an 'empiric-
al' argument: HMM has been mixed with various other filament-
ous cell components, and none of these give a visible arrowhead appearance (or 'decoration') as observed electron microscopically in sections or in negatively stained preparations;
a listing of components which do not give visible reaction includes collagen fibrils, 'tonofibrils', intact microtubules,
microtubule protofilaments, 10 nm filaments, bacterial flag-
ellae, and fibrin (Forer, 1978). Hence, for both these reasons, the arrowhead formation (or 'decoration') is specific for
actin. A further control is to add 1 mM pyrophosphate or 1
mM ATP; both compounds block HMM (or S1) binding to actin
(and remove already bound HMM or S1), and thus both cause the
disappearance of arrowheads (or of decoration).

It should be emphasized that whilst arrowhead formation
is quite specific for identifying actin, it is conceivable
that HMM binds to some or to many other cell components besides actin (see, for example, Heywood et al., 1968, or discussion in Forer, 1974, pp. 323-325). Such binding could
occur and, for reasons discussed above, would not be seen as
arrowhead formation. Hence, even though arrowhead formation
is indeed specific for actin, HMM could interact with and
bind to other cell components besides actin.

It is relevant to discuss other facets of the electron

microscopic study of actin before considering work dealing
with spindle actin. This has recently been reviewed in detail
(Forer, 1978); I herein summarize some of the conclusions relevant to the discussion below. The review (Forer, 1978) can
be consulted for references and for details on all points discussed from here until the next section (Section B-2, on actin
in spindles).

Actin-containing filaments can be seen in myofibrils
either after osmium tetroxide fixation or after glutaraldehyde
fixation. A certain amount of longitudinal shrinkage of each
actin-containing filament is produced by osmium tetroxide
fixation; subsequent ethanol dehydration causes further longitudinal shrinkage of actin-containing filaments in the region
in which these filaments do not interact with myosin (the I-
band), but does not cause shrinkage of such filaments in the
region in which they interact with myosin (the A-band). Thus,
interaction with myosin seems to prevent the ethanol-dehydra-
tion-induced longitudinal shrinkage. The post-osmium longitudinal shrinkage of actin-containing filaments in the I-band is
also prevented by holding the ends of the muscle during dehydration. Lateral shrinkage also occurs, such that in sections
the spacing between actin-containing filaments is only about
70% of that in vivo.

Single actin-containing filaments from muscle are not
straight (as studied using negatively staining), such that,
for example, paracrystalline forms need be used to study the
structure of actin by means of optical diffraction(from electron microscope pictures). Similarly, in sections single actin filaments outside the A-band are rarely seen to be straight
and continuous from A-band to Z-line. Actin-containing filaments are seen more clearly in glycerinated myofibrils than in
non-glycerinated myofibrils, and one can recognize a group of
filaments more readily than one can follow any single filament.

Similar conclusions can be drawn from studies of actin-
containing filaments in non-muscle cells. For example, actin-
containing filaments in non-muscle cells are more readily
identified in groups than when there is only one such filament. (One cannot hold the ends of the filament bundles in
cells as the cells are being dehydrated, or measure lateral
spacing in vivo, as one can with muscle, so there are no
equivalent data on these points in non-muscle systems.) Another example is increased visibility after glycerination.

Actin-containing filaments in non-muscle cells can only be
identified as such by adding HMM in the absence of ATP, and to
do this most workers have utilized glycerol to make the cells
permeable to HMM and to allow the endogenous ATP to leave the
cells. In non-muscle cells, as in myofibrils, glycerination
seems to allow actin-containing filaments to be seen more
clearly than in non-glycerinated cells.

Actin is highly conserved, chemically, from species to
species, as judged by analysis of amino acid sequences. Actin-
containing filaments found in non-muscle cells all can react
with rabbit skeletal muscle HMM, under proper conditions, to
form arrowheads. Nonetheless, while accumulated experience
with muscle proteins can be a 'guide' to planning experiments
on non-muscle actin, one can not rely completely on non-muscle
actin behaving the same as muscle actin, because in many cases
non-muscle actins are chemically different from muscle actins.
As one example, actin from Physarum is antigenically different
from rabbit skeletal muscle actin; indeed, there are at least
three genes that code for different actins in single organisms
(review in Forer, 1978). As further examples, actin-containing filaments in non-muscle cells are different from actin-
containing filaments in skeletal muscle in that some such
filaments are more cold labile and have different salt solu-
bility properties than actin-containing filaments from muscle;
also, monomeric actin in non-muscle cells is sometimes associated with inhibitors that prevent filament formation. Further, in at least two cells (Physarum and human blood platelets)
there are two components which co-electrophorese with actin,
one of which will co-polymerise with the other but will not
polymerise on its own. Finally, a large proportion of the
actin in chick fibroblasts (and chicken brain) will not polymerise under any of a large number of conditions in which
skeletal muscle actin polymerises, even though the chicken
fibroblast actin is not associated with an inhibitor. Nor
will this actin co-polymerise with skeletal muscle actin.
However, this actin can be forced to crystallize by means of
vacuum dialysis, and the crystals can be broken into component
filaments. These filaments look like actin in negatively
stained preparations and bind HMM to form arrowhead complexes,
but when the filaments are dissolved they do not polymerise
into filaments again under conditions in which skeletal muscle
actin polymerises: they can be polymerised only by means of
vacuum dialysis (Bray & Thomas, 1976). Thus there are clear

differences between skeletal muscle actin and at least some of
the non-muscle actins.

Two additional points need be made concerning non-muscle
actin. First, in several non-muscle systems one sees more
filaments (in sections) after reaction with HMM than one sees
(in sections) without HMM treatment. Different explanations
for this seem to apply in different systems. In some systems
HMM causes polymerisation of actin into filaments: prior to
HMM addition the actin existed in non-filamentous form, and
addition of HMM caused polymerisation and formation of arrowhead complexes. In other systems, HMM stabilizes actin-containing filaments against loss during the fixation-dehydra-
tion-embedment procedure: actin-containing filaments are present before fixation but varying numbers (depending on the
system) do not survive fixation and embedment unless they are
complexed with HMM.

Second, several non-filamentous supramolecular forms of
actin exist, including "nets" and "bundles". Neither of these
two forms looks like skeletal muscle actin and neither would
ordinarily be recognized as containing actin. Nor does it
seem likely that either would bind HMM to form arrowhead complexes. Thus these forms of actin are likely to escape detection. In other cells, some supramolecular aggregates of
actin-containing filaments do not bind HMM in situ even after
glycerination and addition of HMM. These filaments will bind
HMM to form arrowhead complexes if they are first splayed out
on a grid, but they do not bind HMM in situ. Thus, with
these known examples of absence of arrowhead complex formation, one needs to be cautious before using negative electron
microscopic data by themselves to rule out the presence of
actin in any given system.

In summary, actin is a universal protein, and actin-containing filaments are found in many cells. Actin-containing
filaments are certainly involved in causing skeletal muscle
contraction, amoeboid movement, cell cleavage, and protoplasmic streaming, and seem to be involved in animal cell locomotion, extension of cellular processes, and other motile phenomena. One generalizes from this and suggests that, until
proven otherwise, one expects actin to be involved in all
cellular motility. Actin-containing filaments are identified
electron microscopically by their reaction with HMM to form
arrowhead complexes, and actin-containing filaments in non-
muscle cells often look like their skeletal muscle counter-

parts. But while one can use the accumulated knowledge of the
chemistry of muscle proteins as a guide to studying non-muscle
actin, there are distinct differences between skeletal muscle
actin and at least some non-muscle actins.

I now consider the question of whether actin is a component of mitotic (and meiotic) spindles.

2. Electron Microscopic Identification of Actin in Spindles: the Evidences

The electron microscopic studies which argue that actin
is indeed a spindle component are discussed in this section,
in which I concentrate on the original data and interpretations. In the next section I discuss critically these evidences, in light of arguments that have been made against these
data and interpretations, and in light of counter-data.

Actin-containing filaments have been found in spindles of
various cell types studied electron microscopically: the reaction with HMM to form arrowhead complexes was used as a
marker to identify actin-containing filaments. The cells in
which spindle actin has been seen in this way are listed in
Table III. In Table III, the term 'microfilaments', which denotes 6-8 nm filaments seen in glutaraldehyde-fixed cells, is
used as shorthand for actin-containing filaments seen in HMM
treated cells. I now discuss these results. (References discussed are those given in Table III unless otherwise specified; the data of Schroeder, 1973, 1976, are discussed separately, in Section B-3, since he argues from his data that
actin is not a spindle component.)

All studies reporting actin-containing filaments in
spindles identified by HMM have used glycerination to make the
cells permeable to HMM (and to allow ATP to leave the cells).
The glycerination technique is a source of contention, to
some, so I summarize details of the techniques used. In all
these studies glycerination was with 50% glycerol in standard
salts solution (standard salts solution is .1M KCl or .05M KCl;
5mM MgCl$_2$; 6mM phosphate buffer, pH around 6.8); glycerination
was at room temperature and glycerination times (before addition of HMM) varied from 3 days (for crane fly spermatocytes)
to 20-30 minutes (mouse neuroblastoma cells, or PtK1 cells),
or even to as little as 1-4 minutes (PtK1 cells). HMM (in 6%
glycerol, or in standard salts solution) was added to the
cells either directly after glycerination, or after the

glycerol was gradually diluted. HMM was sometimes added in the cold (4°C), sometimes at room temperature, and treatment with HMM ranged from 1-2 min (PtK1 cells) to 24-36 hrs (crane fly spermatocytes and locust spermatogonia).

The distribution of actin-containing filaments was studied in various stages of division, or, in neuroblastoma cells, in metaphase only. In anaphase and metaphase, actin-containing filaments between chromosomes and poles were oriented parallel to the microtubules; the one exception was metaphase mouse neuroblastoma cells, in which some filaments were parallel to spindle microtubules whilst most were randomly scattered in amongst the microtubules. Associations between actin-containing filaments and microtubules (or, finding that the 2 elements followed closely parallel, nearby paths) occurred not infrequently in locust spermatogonia, PtK1 cells, and Haemanthus endosperm but did not regularly occur in mouse neuroblastoma cells. Actin-containing filaments were also prominent interzonally, in anaphase, and these were either parallel to the interzonal microtubules (locust spermatagonia), or were both parallel to interzonal microtubules and randomly arranged (crane fly spermatocytes and Haemanthus endosperm cells), or the arrangement was not specified (PtK1 cells). In only PtK1 cells was the orientation and distribution of HMM-binding filaments similar to that of microfilaments in glycerinated cells prior to HMM treatment (see Table III); in PtK1 cells microfilaments were also seen in glutaraldehyde-fixed cells (not-glycerinated), though "less prominently" than in glycerinated cells. In other cells either no microfilaments were seen unless treated with HMM (crane fly spermatocytes), or fewer microfilaments were seen without HMM treatment than with HMM treatment (locust spermatogonia, Haemanthus endosperm), or no comparison was given (mouse neuroblastoma cells).

No clear-cut attachments between actin-containing filaments and chromosomes were illustrated (see Table III); in only locust spermatagonia was such attachment described, but these were not illustrated, nor was serial section analysis described to confirm that such attachments were not artefacts of the large depth of field of the electron microscope. Any such attachment between actin-containing filaments and chromosomes, if they exist, then, would seem to be much less obvious with present methods than those between kinetochore microtubules and kinetochores. More likely, no such attachment

TABLE III

Electron microscopic studies of spindle actin using glycerination
and reaction with HMM (and/or S1): positive results.

Reference	Cells studied	Glutaraldehyde fixed cells MFs?	Glycerinated cells, no HMM (or HMM + ATP, or HMM + pyrophosphate)		Glycerination followed by HMM			
			MFs?	MTs?	MFs?	MTs?	Extra-spindle actin?	MFs attached to chromosomes?
Behnke et al., 1971; Forer & Behnke, 1972b	Crane fly spermatocytes (meiosis)	-	No	No	Yes	Yes or No, depending on the stage of division	Yes	Only possibly, at best
Gawadi, 1971, 1974	Locust spermatocytes (meiosis)	-	Few, only	No	-	-	-	-
	Locust spermatogonia (mitosis)	-	Yes ("in small numbers.")	Yes	Yes (more than without HMM)	Yes	Yes	Some seen directly attached, but not necessarily at kinetochores
Schroeder, 1973, 1976	HeLa cells	-	Yes (few)	Yes	Yes (few; never in bundles)	Yes	Yes	No

Hinkley & Telser, 1974	Mouse neuroblastoma cells (mitosis)	-	Yes	Yes	Yes	Yes	Filaments are rarely near chromosomes: only possible contact, at best
Schloss et al., 1977	Rat Kangaroo (PtK1) cells (mitosis)	Yes	Yes ("more prominent..." than with no glycerination)	Yes (with orientation and distribution similar to glycerination only)	Yes	Yes	None reported, so presumably none were seen to
	Rat embryo cells in culture (mitosis)	-	Yes	Yes	-	-	-
Forer & Jackson, 1975, 1976	Haemanthus (African blood lily) endosperm cells (mitosis)	-	Some	Yes (in large numbers, more than without HMM)	Yes	No (very few such filaments were seen)	None seen (serial sections were studied)

MFs = microfilaments

MTs = microtubules

- means that no data were reported on these points

45

exists. Related to this is the question of the polarities of
the spindle actin, i.e., the direction of the arrowheads on
the actin-containing filaments. In muscle, the actin-containing filaments have arrowheads which point away from the Z-line
and towards the myosin, in the A-band, as in Fig. 3. By analogy, one would expect that if actin was attached to a chromosome at the kinetochore the chromosome would be a Z-line
equivalent, and arrowheads would point towards the pole (Fig.
4). Or, conversely, if the actin was not attached to the
kinetochore but rather to something else then the arrowheads
should point towards the chromosome (Fig. 5), in which case
the chromosome or associated microtubules (or other components) are analogous to the myosin. Arrowhead directions are
seen only rarely in sectioned material (review in Forer,1978),
and with respect to spindle actin only 2 articles have given
data on actin polarity: (a) in locust spermatagonia (Gawadi,
1974), actin-containing filaments between chromosomes and
poles in metaphase (and oriented parallel to spindle microtubules) had arrows pointing predominantly towards the chromosomes (45 out of 62); (b) in Haemanthus endosperm (Forer &
Jackson, 1976) only 11 clear arrowheads were seen in the
chromosome-to-pole region, out of "hundreds" of filaments
photographed in metaphase and anaphase, and 8 out of 11 arrows
pointed towards the chromosomes. A further article has emphasized the rarity of arrowheads: in PtKl cells (Schloss et al.,
1977), very few arrowheads clear enough to get directionality
were seen in 60 spindles studied. Thus, though the arrowheads
would seem to point towards the chromosomes, in agreement with
the lack of attachment of actin-containing filaments to the
kinetochore, the difficulty in attaching meaning to the polarities is that only a very small percentage of the filaments
have clear polarities in electron micrographs.

   While perhaps not directly relevant to the discussion at
hand, it is clear from the data summarized in Table III that
microtubules in meiotic spindles seem to react differently to
glycerination than do microtubules in mitotic spindles: in
glycerinated meiotic cells (crane fly spermatocytes, locust
spermatocytes) one does not see microtubules; in glycerinated
mitotic cells (the remainder in Table III) one does. While
the data so far are rather limited, this does not seem to be
simply a technical difference, because microtubules were seen
in glycerinated locust spermatagonia in the same testis lobes
as the spermatocytes. That microtubules are seen after HMM

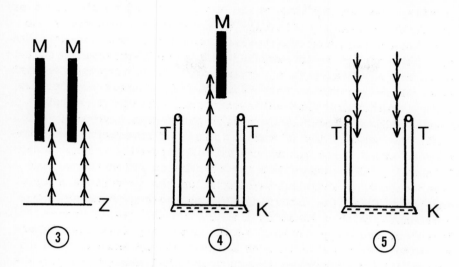

Fig. 3.   Illustrates the arrangement in skeletal muscle:
the Z-line (Z) is the site of attachment of actin-containing
filaments, that have polarities (arrows) pointing towards the
thick myosin filaments (M).   Figs. 4 and 5 represent possible
attachments (and polarities) of actin-containing filaments in
spindles.   In Fig. 4, actin-containing filaments are attached
to the kinetochore (K), as are microtubules (T), with actin
polarity (arrows) pointing towards the myosin filament (M)
(that may or may not exist in spindles).   In Fig. 5 the actin-
containing filament is anchored somewhere in the spindle, with
polarity (arrows) towards the kinetochore (K) and kinetochore
microtubules (T).

treatment of meiotic cells whereas none were seen without HMM
treatment suggests strongly that actin interacts with these
microtubules, as discussed in detail previously (Forer, 1974).
    It is relevant to later discussion to note that, unlike
the situation in the animal cells studied, there was very
little extra-spindle actin seen in Haemanthus endosperm cells.
This finding has been confirmed by serial section reconstruction analysis of other Haemanthus endosperm cells (Forer &
Jackson, in preparation), glycerinated in various other ways
(e.g., using 25% glycerol with 0.1% Triton-X-100 added).
    Summarising these data on HMM-binding filaments in the

spindle, one can say that while actin is found in the spindle, the actin-containing filaments do not seem to be attached to chromosomes, and the qualitative observations made to date do not allow one to decide whether or not actin-containing filaments are associated in a regular way with chromosomes or with kinetochore-associated microtubules. All authors nonetheless argue, cautiously, that actin-containing filaments probably have a role in producing force for chromosome movement. All admit the possibility of various kinds of artefacts, but nonetheless argue for a functional role for spindle actin. Forer & Behnke (1972b) argue that the changes in orientations and arrangements of actin-containing filaments in various stages suggest that actin-containing filaments are genuine spindle components, with a functional role. Gawadi (1974) argues likewise, and suggests that actin-containing filaments interact with microtubules in such a way that, by analogy with muscle, the microtubules are surrogate Z-lines (Fig. 5): actin-containing filaments "terminate" at microtubules (i.e., they interact with microtubules in a fixed positional relationship). As a corollary, she suggests that chromosomes are surrogate myosins. Forer & Jackson (1976) suggest that actin-containing filaments are on the outside of microtubule bundles, and have some functional role. Schloss et al. (1977) suggest that there is an intimate interaction between microfilaments (actin-containing filaments) and microtubules, with the microfilaments being closely applied to the microtubule wall and thereby only rarely being seen in glutaraldehyde-fixed (non-glycerinated) cells.

Before considering objections to these data, and before considering counter-data, it is relevant to consider electron microscopy of glutaraldehyde-fixed (not-glycerinated) cells, because 6-8 nm (diameter) microfilaments have been reported in spindles in various cells. Ryser (1970) described the regular occurrence of large numbers of microfilaments in the intranuclear spindle of Physarum, in all stages of mitosis; these microfilaments looked like cytoplasmic actin filaments, and Ryser (1970) argued that these were spindle actin filaments that were probably involved in generating force for chromosome movements. Lewis et al. (1976) described intranuclear microfilaments in Paramecium micronuclei during mitosis, but the documentation of the presence of microfilaments was not compelling. Müller (1972) described microfilament bundles seen in early-prometaphase and mid-prometaphase spindles in crane fly spermatocytes, and Forer & Brinkley (1977) described single microfilaments in mid-anaphase crane fly spermatocyte spindles, near microtubules. In Haemanthus endosperm cells Bajer & Molè-Bajer (1969) found "frequently in prophase and later stages a continuous transition between MTs [microtubules] and bundles of very thin fibrils..." the size of microfilaments. Goode (1975) found microfilaments in embryonic chick heart muscle cells in mitosis, but only a few were seen and, to quote Goode (1975), there was "no ultrastructural evidence of a specific regular interaction of mitotic apparatus microtubules and any other filament type... at any stage of mitosis." On the other hand, microfilaments were consistently seen in spindles of Xenopus heart cells in primary tissue culture, presumably fibroblasts (Euteneuer et al., 1977). Individual microfilaments were seen rather than bundles, and these were scattered in amongst the spindle microtubules and oriented primarily parallel to the microtubules; these filaments were seen in all stages of division, and the authors argue that in view of the regular orientation and distribution of the filaments they are probably involved in the function of the spindle. Microfilaments were seen in the stem body of human cells in culture (WI-38 and HeLa cells), but these were not seen in metaphase or anaphase spindles (McIntosh & Landis, 1971). In PtK1 cells, McIntosh et al. (1975) identified microfilaments in cross-sectioned spindles (presumably in serial-sections), quite near to microtubules, and Schloss et al. (1977) described microfilaments in longitudinally-sectioned spindles, also quite close to microtubules.

There are difficulties in interpreting the above observations however, in that, for example, the distribution of these filaments is not known, so they may not be regularly associated with chromosomal spindle fibres. Further, all 6-8 nm filaments do not necessarily contain actin. Positive identification of actin requires interaction with HMM, or other marker. One would need to describe the positions of these microfilaments in spindles in glutaraldehyde-fixed (non-treated) cells and show that those filaments which react with HMM have the same distribution as the filaments found in glutaraldehyde-fixed (not-glycerinated) cells, if one wished to use these observations to prove that actin-containing filaments exist in these spindles in vivo. Nonetheless, that microfilaments are seen in glutaraldehyde-fixed (not-treated) Haemanthus endosperm cells, PtK1 cells, and crane fly spermatocytes,

the same cells in which actin-containing filaments (which react with HMM) are found (Table III) is encouraging to those
like myself who believe that actin-containing filaments are
involved in moving chromosomes. I now consider arguments and
data that are counter to the view that actin-containing filaments are functional spindle components.

3. Electron Microscopic Identification of Actin in Spindles:
    Counter-arguments and Counter-evidences

    Porter (1973) has discounted the evidence identifying
actin-containing filaments in the spindle; to quote him, this
is "because I suspect contamination in experiments which use
heavy meromyosin to detect actin". There are two interpretations of this statement. One of these can be ruled out immediately, and that is that the HMM is contaminated with actin-
containing filaments. One does not see actin-containing filaments in electron microscopical observation of negatively
stained HMM preparations using the same HMM as used to identify spindle actin (e.g. Forer & Behnke, 1972a, 1972b); contamination with actin-containing filaments is readily seen under
such conditions, when HMM preparations are contaminated (e.g.,
Forer, 1978), and hence one concludes that this objection is
not valid. It is possible, however, that this statement really refers to "contamination" of the spindle by extra-spindle
actin; this argument is discussed below.
    Arguments against the presence of actin-containing fila-
ments in crane fly spermatocyte spindles were presented by
LaFountain (1974, 1975). Crane fly spermatocytes were not
treated with HMM but were directly fixed with glutaraldehyde
and then post-fixed and processed in the usual way (LaFountain, 1974), or were fixed with glutaraldehyde-tannic acid and
then post-fixed and processed in the usual way (LaFountain,
1975). Microfilaments (5-7 nm in diameter) were seen in
cleavage furrows, oriented equatorially, in both glutaraldehyde and glutaraldehyde-tannic acid fixed cells, and none were
seen in other regions of the cortex in metaphase and anaphase
(LaFountain, 1974). No microfilaments were seen in the
spindle, with either fixative. LaFountain (1974, 1975) argued that spindle microfilaments found after HMM treatment do
not really exist in spindles in vivo, because, to quote La-
Fountain (1974), "Since cortical filaments have been preserved
by our techniques, one would expect that at least some micro-

filaments in spindles would have been preserved and detected, provided they are present in the spindle"; hence, he questions "whether spindle microfilaments actually exist in vivo." Where then did the actin-containing filaments come from that were seen in crane fly spermatocyte spindles after glycerination and HMM treatment? LaFountain (1975) speculated that perhaps "glycerinated spindles are "contaminated" with actin...", which originally was outside the spindle. Or, that microtubules bind HMM to give arrowhead complexes as the microtubules are broken down by the glycerination. Or that the actin exists in spindles in the globular (non-polymerised) form, and, presumably, polymerises during the glycerination-HMM treatment.

There are several flaws in the LaFountain (1974, 1975) argument. First, one suggestion can be eliminated immediately: microtubule protofilaments do not bind HMM in any visible manner (review in Forer & Behnke, 1972a), and thus the spindle 'decorated filaments' are not simply microtubule protofilaments, and are actin-containing filaments, as reviewed above. Also, in other cell types microtubules exist in the same spindles as do HMM-binding filaments (Table III). Second, all cortical filaments have not been preserved by LaFountain (1974): actin-containing filaments are seen (after glycerination and HMM treatment) around the entire circumference of crane fly spermatocytes, in large numbers, in regular orientations, and in all stages of meiosis (Forer & Behnke, 1972c). These were not seen by LaFountain (1974, 1975). Similar prominent bundles of actin-containing filaments are seen in cortices of many animal cells (review in Holtzer et al., 1973), so this contradiction is not due to the observations of one set of workers only. LaFountain's original argument was: all cortical microfilaments are preserved and seen, spindle microfilaments are not seen, therefore spindle microfilaments do not exist in vivo. But since the prominent and regularly arranged bundles of microfilaments in the cortex outside the cleavage furrow were not seen, this argument reduces to: cleavage furrow microfilaments are seen, but other cortical microfilaments are not and neither are spindle microfilaments. That is to say, LaFountain missed one prominent class of actin-containing filaments and could have missed as well the less prominent microfilaments in the spindle. Hence this argument is not compelling. Third, microfilaments have been seen in non-treated glutaraldehyde-fixed cells, as described above, including spermatocytes of the same species of crane fly studied by LaFountain (see Forer & Brinkley, 1977). Indeed, despite LaFountain's claims to the contrary, possible longitudinal-section images of microfilaments can be seen in Fig. 1 of LaFountain (1975) which look like the 'microfilaments' illustrated in Forer & Brinkley (1977) and in Schloss et al.(1977). Likewise, there are cross-sectional images in the spindles (from glutaraldehyde-tannic acid fixed cells) which look like cross-sectioned microfilaments: the profile at 2.8 cm.from the left and 1.1 cm. from the bottom of Fig. 2b of LaFountain (1975) looks exactly like cross-sectioned actin-containing filaments from muscle illustrated in the inset to Fig. 2a of LaFountain (1975). The final point is the argument that spindles are "contaminated" with extra-spindle actin: this is discussed below. In sum, the arguments of LaFountain (1974, 1975) are quite unconvincing, except possibly for the speculation that spindle actin might be "contamination" from without.

Physiological lines of evidence have been used to argue that actin-myosin interactions are not involved in mitosis. One such argument is based on studies of the isolated mitotic apparatus (MA). Sakai et al. (1975) reported that chromosome movement could be induced in MA isolated in glycerol, if the MA were suspended in tubulin, in microtubule polymerising medium plus ATP. Sakai et al. (1976) extended these observations to test the nucleotide specificity of the movement and to test the effects of antibodies against dynein and of antibodies against myosin. They found that chromosome movement required ATP and that other nucleotides could not substitute for the ATP. They found, further, that antibodies against flagellar dynein (the flagellar ATPase) blocked chromosome movement in isolated MA whilst antibodies against egg myosin (from the same species from which MA were isolated) had no effect on chromosome movement. Therefore, as argued by Mohri et al. (1976), Sakai et al. (1976), and Mabuchi & Okuno (1977), chromosome movement is most likely due to dynein ATPase interacting with microtubules and is not likely to be due to myosin interactions with actin.

There are several flaws in this argument, though. First is the question of how well the movement in MA in vitro represents chromosome movement in vivo. Anaphase in sea urchin zygotes is reasonably fast, and usually lasts no more than 10 minutes in vivo; during this time the chromosomes move polewards 10-15 microns, at velocities, therefore, of about 1

micron per minute. In the in vitro movement, on the other
hand, in isolated sea urchin zygote MA, the chromosome to pole
distances shortened by only 2-3 microns; this shortening took
place in 1-2 hours, so the in vitro chromosome velocities
measured by Sakai et al. (1976) are about 1.5 microns per
hour, or about 60 times slower than in vivo. Also, chromosomes in vitro moved only a fraction of the distance moved in
vivo (3 out of 10 microns). Thus, for both these reasons, the
chromosome movement studied in vitro is quite different from
in vivo movement, and the relevance of the antibody experiments to the in vivo situation is not clear. Second is the
question of how accurate the measurements are. In general one
needs to consider resolution limits of the microscope and consequent accuracy of measurements, but not enough experimental
information is given in the paper by Sakai et al. (1976) to
judge these points. Nonetheless, inspection of the published
curves leads one to question the accuracy of the measurements,
for the following reason. In the published curves, Sakai et
al. (1976) give measurements of pole-to-pole distances (spindle lengths), chromosome-to-pole distances, and interzonal
distances (distances between the separating chromosomes).
Clearly, the spindle lengths should equal the interzonal distances plus two times the chromosome-to-pole distances, within the accuracy of the measurements. The correspondence between these numbers, then, is a measure of the accuracy of the
measurements. If one inspects the curves, one finds many instances of 2 to 4 micron differences between spindle lengths
and the sum of interzonal distances plus two times the chromosome-to-pole distances: examples include Figs. 1b, 2b, 3a, 3b,
3c, and 5b of Sakai et al (1976). Therefore, since individual
curves are inaccurate to within 2-4 microns, and since the
total movement (2 times the shortening of the chromosome-to-
pole distances) is only 4-6 microns, one questions how much
movement indeed has occurred. The final points are missing
controls: would the antibody against myosin block the contractile function of myosin in muscle (or other myosin system)
under the conditions of the experiment? No such experiments
were reported, but are necessary, because McIntosh et al.
(1975) were unable to block the contraction of glycerinated
muscle by application of antibodies against myosin or of antibodies against actin. Also, one needs a control to ensure
that the antibody against dynein does not affect chromosome
movement indirectly via an effect on the tubulin polymerisa-

tion medium in which the MA need be suspended. Thus, in sum,
the observations are of inadequate accuracy and have question-
able relevance to postulated roles of myosin involvement in
chromosome movement.

Another physiological experiment has been used to argue
against actin-myosin interactions being functional in mitosis;
in this experiment antibody against myosin was injected into
blastomeres (at the 2 cell stage) at varying times before cell
cleavage (Mabuchi & Okuno, 1977). Sakai et al. (1976) have
argued from these experiments in the following way. Injection
of antibody blocked cleavage, but did not block chromosome
motion. The antibody does function after being injected into
the cell, since cleavage is blocked; therefore, they argue,
"the fact that no inhibition of chromosome motion was caused
by the anti-egg myosin γ-globulin provides evidence that the
chromosome motion ... is not dependent upon a myosin-actin
system ..." (Sakai et al., 1976). (Mabuchi & Okuno, 1977, who
presented the relevant data, do not make so explicit an argument.) One must examine the experiments in detail in order to
critically assess this argument, so I first summarize the relevant experimental details.

Mabuchi & Okuno (1977) prepared antibodies against myosin
isolated from starfish eggs. The antibodies had no effect on
myosin Ca-ATPase activity, nor on myosin Mg-ATPase activity,
but did block the actin-activated Mg-ATPase activity of myosin. Mabuchi and Okuno injected antibodies into single blastomeres of starfish embryos at the 2-cell stage: they injected
different doses of antibody and injected at various times in
the cell cycle, namely in interphase (before nuclear membrane
breakdown), after nuclear membrane breakdown, and just at the
start of cleavage. The cells took 15 minutes to progress from
nuclear membrane breakdown to cleavage, and the injected antibody was spread throughout the cell in less than one minute
(as measured using fluorescently labelled non-immune serum);
thus, they argued, diffusion was not a limiting factor in the
experiments. The results were as follows. In injections
prior to nuclear membrane breakdown, 0.3 ng of antibody blocked cleavage in 100% of the cells (0.2 ng blocked cleavage in
only 50% of the cells). In injections after nuclear membrane
breakdown, however, 0.5 ng of antibody, almost twice the previous concentration, inhibited only 50% of the cell cleavages.
Injections at the onset of cleavage did not block cleavage at
all: after injection of 0.5 ng of antibody there was furrow

regression in 2 out of 9 cells, whereas after injection of
0.5 ng of pre-immune gamma-globulin, in the controls, there
was blockage of cleavage in 1 out of 8 cells. If one accepts
that cell cleavage is a result of actin-myosin interactions in
the cleavage furrow ('contractile ring'), as Mabuchi & Okuno
(1977) do, then one concludes from this aspect of the experiment that the antibody against myosin blocks the setting up of
a functional contractile ring, and hence blocks cleavage, but
that the antibody does not block the actin-myosin interactions
once the contractile ring is already set up.

How about the supposed lack of effect on chromosome movement? This was studied in 17 cases in which antiserum was injected prior to nuclear membrane breakdown, in all of which
cleavage was blocked. (No doses were reported.) In 5 cases
no MA were formed; in the remaining 12 cases the MA that were
formed were, to quote Mabuchi & Okuno (1977), "small and obscure". (There was no documentation of presence or absence of
MA, or of altered MA, or even mention of which microscope system was used for these observations, however.) Did the "small
and obscure" MA function normally? As far as I can determine
from the published work (Mabuchi & Okuno, 1977), chromosome
movement was neither seen nor measured; the only relevant data
presented were that "formation of daughter nuclei was observed
in 9 out of the 12 cells..." (Mabuchi & Okuno, 1977). Presumably the argument regarding lack of effect on chromosome movement derives from the observation that 2 daughter nuclei were
formed in each of these 9 cells. I now consider these experiments with regard to the conclusion of Sakai et al. (1976)
that chromosome movements in mitosis are "not dependent upon
a myosin-actin system ...".

Several assumptions inherent in the experimental approach
should be considered before considering details of the results. The first point is the validity of experiments in
which non-immune serum is used to estimate the diffusion of
specific antiserum (in this case, antiserum containing antibodies against myosin). Experimentally, Mabuchi & Okuno
(1977) observed that non-immune serum freely diffused throughout the cell, in less than one minute, and they assumed,
therefore, that antibodies against myosin would do likewise.
But this assumption is not necessarily valid. For example,
there is a large amount of non-diffusible myosin in the cell
cortex (Mabuchi, 1973, 1974, 1976), and if there are fewer
antibody molecules injected into the cell than there are

myosin molecules, high affinity antibodies would be sequester-
ed at the positions of the myosin and no antibody molecules
would be free to diffuse elsewhere in the cell. [It is relevant to note that antibody against dynein reacts in exactly
this way: antibodies against dynein blocked the beating of
sperm, and after washing away excess antibody and immersing
the sperm in solution not containing antibodies, the antibodies against dynein remained bound to the sperm, and the
sperm were not reactivated (B. Gibbons et al., 1976).] One
therefore needs to do a different control. One such control
would be to add fluorescently-labelled antibodies against myosin, and to see whether such antibodies are indeed sequester-
ed and to see if antibodies are present in the spindle (or
other region of interest). Another possibility is to demonstrate chemically that the antibodies are in large excess over
intracellular myosin, and that not enough additional myosin is
synthesized after the antibodies are added to deplete the
antibody pool.

A second point is the assumption that antibodies directed
against myosin ATPase will block contractile processes involving myosin. That is to say, it is assumed that if cell cleavage involves myosin interactions with actin and if chromosome
movement (or anything else) involves myosin interactions with
actin, then addition of antibodies that are against myosin
ATPase will block (stop) cell cleavage, chromosome movement,
or anything else that requires myosin interactions with actin.
But this assumption is not valid: several experiments demonstrate that antibody directed against myosin might completely
block actin-myosin interactions if the antibodies are added
with myosin before actin and myosin are able to interact, but
that the antibodies do not completely block actin-myosin interactions if the antibodies are added after actin and myosin
are able to interact. For example, antibodies against muscle
myosin blocked a large proportion of the myosin Ca^{++}-ATPase
and Mg^{++}-ATPase activity, they blocked a large proportion of
the actin-activated Mg^{++}-ATPase, and they blocked actomyosin
superprecipitation, but they did so only if the myosin reacted
with the antibodies prior to interaction with actin: there was
no effect when antibodies were added directly to actomyosin
(Puszkin et al.,1975). In another example, McIntosh et al.
(1975) were unable to block the contraction of glycerinated
muscle by addition of antibodies against myosin or against
actin. In a further example, antibodies against Physarum

myosin (Kessler et al., 1976) greatly inhibited the myosin
Ca^{++}-ATPase, but antibodies that were added to actomyosin did
not block superprecipitation: superprecipitation was delayed
(the amount of delay depended on the amount of antibody), but
the superprecipitation itself was not blocked, nor was there
much effect on the extent of superprecipitation, as judged by
change in absorbance (Nachmias & Kessler, 1976). Thus, in any
experiments on the effects of antibodies on chromosome movement it is necessary to measure chromosome velocities, and to
determine if the start of anaphase is delayed, for, by analogy with these experiments, antibodies might cause chromosome
movement to be delayed, or slowed, without causing movement to
be completely blocked. In yet another example, antibodies
against myosin blocked cleavage only when they were injected
into cells much prior to the time the contractile ring was set
up (Mabuchi & Okuno, 1976), as discussed previously. As a
final example it is relevant to consider the effect of antibodies against dynein-ATPase on the movement of sperm (B. Gibbons et al., 1976; I. Gibbons et al., 1976). Addition of
antibodies to reactivated (beating) sperm resulted in gradual
decrease of the beat frequency; the beat frequency declined
rapidly in the first 5 - 15 minutes, and then more slowly up
to 60 minutes, but even after 60 minutes the sperm continued
to beat, albeit with a greatly reduced frequency. The amount
of reduction in beat frequency depended on the amount of antibody added. These results emphasize the point made above,
that chromosome velocities need be measured, for, by analogy
with the results on sperm, one might even expect slowing down
of movement rather than cessation. Also relevant is the comparison between adding high concentrations of antiserum to
"rigor" sperm vs. to reactivated, motile sperm. When high
concentrations of antibodies against dynein-ATPase are added
to reactivated sperm the beat frequency is greatly reduced,
but sperm continue to beat. But when antibodies are incubated
with rigor sperm for 30 sec. prior to addition of reactivation
medium, then the sperm do not beat (B. Gibbons et al., 1976).
Here, too, then, function which is blocked completely by in-
cubation of antibodies with the separate (non-functional) component is not blocked when added to the same component which
is in a functional state.

  A third point to consider is whether myosin in the spindle would be the same as myosin in the cortex, and whether
antibodies against the active site of one myosin would necess-

arily block the active site of the other myosin. Myosins from
different tissues do not cross-react with antibodies against
myosins of the other tissues (e.g., Bruggmann & Jenny, 1975;
Puszkin et al.,1975; Fujiwara & Pollard, 1977; Pollard et al.,
1976, 1977), and indeed, antibodies against the ATPase of
dynein-1 from sperm do not inhibit the ATPase of dynein-2 (B.
Gibbons et al., 1976), so it is not unreasonable to expect
that different myosins in the same cell would not be blocked
by antibodies to only one of the types of myosin. Nor is it
unreasonable to expect that there are different myosins in the
same cell, for indeed 3 different actins occur in single cell
types (review in Forer, 1978), and recent evidence indicates
that there may be as many as 29 different genes coding for
actins (Tobin & Laird, 1977; and S.L. Tobin, personal communication). Hence the possibility that the injected antibody
would not even react with spindle myosin (if it exists).

The final point is that even if one could prove definitively that there was no myosin in spindles, this does not
rule out a functional role for actin, for, as Tilney has emphasized (e.g., Tilney, 1975), some motile systems utilize
actin, but without myosin.

I now consider details of the experiments of Mabuchi &
Okuno (1977). First, as discussed above, the antibodies
blocked cleavage only if injected much earlier than cleavage,
so, in agreement with other work discussed above, one assumes
that the antibodies do not block a functioning system once it
is set up but that the antibodies will block the setting up of
a functional system. [Mabuchi & Okuno (1977) give data that
their antibodies inhibit the actin-activated Mg^{++}-ATPase of
myosin, which might be considered as a counter-argument; but
Mabuchi & Okuno (1977) do not state whether this was done by
pre-incubating the myosin with antibody, as is common (e.g.,
Trenchev & Holborow, 1976; Gröschel-Stewart et al., 1977), or
whether the antibody was added after the actin and myosin were
already mixed.] One could argue from the Mabuchi & Okuno
(1977) data, therefore, that myosin is a spindle component:
since the antibodies block formation of a functional myosin
structure, and since either no MA or "small and obscure" MA
were formed after injection of antibodies, one could conclude
that there is spindle myosin, and this myosin functions in
setting up normal-sized MA. As a second point, even if one
uses the mere presence of daughter nuclei as criteria for the
occurrence of chromosome movement, Mabuchi & Okuno (1977) show

that chromosome movements never occurred in almost half the
cells (in 5 cells no MA were formed, and in 3 cells no daughter nuclei were found, for a total of 8 out of 17); this is
quite different from the interpretation "no inhibition of
chromosome motion". Further, in the other cells chromosome
movements were never measured, and thus they could have been
slower, or anaphase could have been delayed.

I conclude that the data of Mabuchi & Okuno (1977) are
not really relevant to the question of myosin involvement in
chromosome movement, because of lack of data dealing with the
basic assumptions discussed above (namely, diffusion of antiserum vs non-immune serum, cessation of movement vs slowing or
delaying of movement), and that even accepting their interpretation spindle myosin may not be inhibited by their antibodies, or spindle actin may function without myosin. Further,
their data could even be used to argue for the presence and
function of spindle myosin.

Injection of antibodies against myosin were also reported
to block cleavage but not chromosome movements by Kiehart et
al. (1977). These experiments seem to rule out some objections I raised against those of Mabuchi & Okuno (1977), but
only when detailed descriptions of the experiments are available will one be able to critically evaluate the results.

Schroeder (1973, 1976) has argued on several grounds that
actin has no role in chromosome movement. For one, Schroeder
(1973, 1976) studied HeLa cells after glycerination and HMM
treatment; he found some actin-containing filaments in these
spindles, but not "in functionally significant numbers..."
(Schroeder, 1973). In metaphase spindles, between chromosomes
and poles, "A few - but only a few - solitary filaments can be
seen ...", and Schroeder has "never seen bundles of filaments
in this region..." (Schroeder, 1976). In anaphase, actin-containing filaments are seen interzonally, and in the cell cortex, but "are practically never seen in chromosome-to-pole
regions" (Schroeder, 1976). Second, not only are there only
"a few" actin-containing filaments, the distribution of these
filaments is wrong: the filaments "either remain fixed in
place during chromosome motion or are preferentially located
behind the moving chromosomes..." (Schroeder, 1976), both of
which seem "inconsistent" with actin associations with a
traction fibre. Third, Schroeder (1976) studied mitoses in
7-day chicken embryo gizzard cells, in which the smooth muscle
contractile filaments were already present. Schroeder saw

solitary filaments in the spindles, that looked like the
muscle filaments, but "there is no preferential concentration,
alignment or association of these putative actin microfilaments in any portion of the MA at any stage of division..."
(Schroeder, 1976). This is "additional evidence against a
significant role for actin in chromosome motion..." (Schroeder, 1976). The final argument is that there are "sharp con-
tradictions" between the results of electron microscopic detection of actin in spindles and the results of fluorescent
microscopic detection of actin in spindles, and therefore "the
data are seriously contradictory and tend to nullify each other..." (Schroeder, 1976). I discuss this last point in Section IV-D, after considering in detail the methods and results
of fluorescence microscope detection of actin; I now consider
the first three points raised by Schroeder.

First, in considering the absence of "functionally significant numbers of filaments", the wrong distribution of filaments, and the absence of filaments in spindles of glutaraldehyde-fixed (non-treated) cells, one should point out that, as
summarized above, others have studied similar cells, both before and after glycerination and HMM treatment, and have given
different descriptions of actin-containing filaments in spindles (e.g., Euteneuer et al., 1977; Schloss et al., 1977).
Part of the difficulty in resolving this apparent contradiction in results is in the question of what one expects to find,
and is in the lack of quantitative data from either Schroeder
or those like myself who claim that actin in the spindle is
present in proper amounts, orientations, and distributions to
be functional (Table III). To consider one point, Schroeder
argues that in HeLa cells there are only a "few" actin-containing filaments, in numbers that are not "functionally significant". The data summarized in Section II of this chapter
show that only 1 actin filament may be required per chromosomal spindle fibre, so that even "a few" may be functionally
significant. For example, HeLa cell spindles have 1000 - 1500
microtubules (e.g., McIntosh & Landis, 1971), so that 100 actin-containing filaments would be "a few" by comparison, yet
might be more than enough to do the job, if they were distributed properly. Nor is there any a priori reason to expect
bundles of filaments. I can only conclude that this disagreement on whether actin-containing filaments in the spindle may
be functional can not be resolved without quantitative data on
the distribution of actin in the spindle, primarily to see if

there are actin-containing filaments associated with each and
every chromosome or chromosomal spindle fibre; this is the re-
quirement of any force-producer, and if actin-containing filaments are not so distributed, then the actin-containing filaments can not be functionally involved in causing chromosome
movement. Until we have these data, the disagreement is essentially unresolvable.

Second, it is relevant to consider another aspect of this
disagreement, and that is the difficulty in seeing microfilaments in non-treated cells. Schroeder has made his expectations clear: to quote him (Schroeder, 1976), "I would expect
that if actin were involved in force production (in the spindle) that its location would be easily and reproducibly vis-
ualized". This expectation has indeed not been fulfilled; but
this expectation need not fit reality. McIntosh et al.(1976),
Sanger & Sanger (1976), Cande et al. (1977), and McDonald et
al. (1977), for example, have emphasized the problem of fixation of actin filaments and suggest that, perhaps like spindle
microtubules that were rarely seen prior to the advent of
glutaraldehyde fixation, spindle microfilaments are just not
well preserved with present methods of fixation. The literature on fixation of actin-containing filaments in muscle, discussed above, emphasizes this point (review in Forer, 1978).
On the other hand, Schloss et al. (1977) suggest another possibility: they suggest that microfilaments are not seen in PtKl
cell spindles because they are obscured by closely apposed
microtubules. Schloss et al. (1977) document this in PtKl
cell spindles, and recent work by Jackson and me (in preparation) on serially-sectioned spindles in Haemanthus endosperm
cells confirms such closeness between microfilaments and
microtubules. So, for these and other reasons, that Schroeder's expectation is not fulfilled does not mean that actin
does not function in spindles.

Two other counter-arguments remain to be discussed. One
is that though actin is seen in the spindle, "the possibility
that the actin was, in life, located elsewhere in the cell has
certainly not been excluded..." (Nicklas, 1975); this argument, that actin "relocates" during glycerination and HMM-
treatment, has also been presented by Porter (1973) and
LaFountain (1975), as discussed previously. The second is
that actin is "trapped" in the spindle, and has no real role:
to quote Sato et al. (1976), "The spindle is ... formed at
each mitosis in a cytoplasmic environment that contains a

quantity of actin and some actin filaments. Thus it is not surprising to find some actin-like filaments both inside and outside the spindle...".

Forer & Jackson (1976) have tried to deal with the relocation problem by studying spindles in Haemanthus endosperm cells: after glycerination and heavy meromyosin treatment there were very few actin-containing filaments outside the spindle and relatively large numbers inside the spindle. Thus there was no large extra-spindle source of actin-containing filaments which could "contaminate"' the spindle. Jackson and I have subsequently studied (and photographed) Haemanthus endosperm cells undergoing glycerination, and we have found that cells glycerinated as in Forer & Jackson (1976) shrink considerably when they first are placed in 50% glycerol, but that after 3 minutes or so the cells suddenly expand, to near original size, whilst the chromosomes maintain positions near those seen originally. Thus actin could "relocate" during the shrinkage period. However, cells prepared in different ways had different degrees of shrinkage and swelling, yet all had the same distribution of actin. This argues against "relocation", but some change in shape has been observed during all treatments studied to date so we can not definitively rule out "relocation" of actin-containing filaments during the procedures.

Could actin be "trapped" in the spindle? Actin-containing filaments found in intranuclear spindles (Ryser, 1970; Lewis et al., 1976) would rule out this objection (and would rule out the "relocation from the cytoplasm" objection), except that these filaments have not been demonstrated (by HMM-binding) to contain actin. And even if so, this does not necessarily mean that such filaments would function in chromosome movement. A key to deciding whether actin might function in causing chromosome movement is to see if actin is associated with each and every chromosomal spindle fibre and/or chromosome. Such a consistent finding would also argue against "relocation" and "trapping", since these would predict random arrangements of actin.

To see if there is such a consistent arrangement of actin-containing filaments in spindles, Bill Jackson and I have begun serial-section reconstruction of actin-containing filaments in Haemanthus endosperm cells, and I summarize briefly some initial findings. First, as noted above, we confirm that there is little extra-spindle actin. Second, we confirm

exactly the description by Schloss et al. (1977) regarding intimate associations between actin-containing filaments and
microtubules. Throughout the spindle, actin-containing filaments and microtubules are in very close association; the two
are often so close together that microtubules obscure the actin-containing filaments, such that, as in Schloss et al.
(1977), actin-containing filaments are seen to enter the sections just as microtubules leave the sections. The same situation applies with the same microtubules and actin-containing
filaments through several serial sections, and over distances
of microns. (Some such associations between actin-containing
filaments and microtubules are illustrated in Figs. 6 - 8.)
Third, we have so far analyzed longitudinal serial sections
through 4 kinetochores (and associated microtubules), in one
metaphase cell. We have found several actin-containing filaments associated with each group of kinetochore microtubules;
the filaments were in amongst the microtubules as well as on
the outside of the bundle, and were found from near the kinetochore to at least several microns from the kinetochores towards the poles (Forer & Jackson, in preparation). At least
preliminarily, then, these findings support the view that actin-containing filaments in the spindle function in causing
chromosome movement.

In conclusion, I have summarized various arguments raised
against the possibility that actin-containing filaments function in mitosis. Many of these arguments were dismissed, but
one is left with the 3 counter-arguments that actin "relocates" during glycerination, or is "trapped" in the spindle, or
just does not function in causing chromosome movement. There
are arguments against "relocation" from the cytoplasm, since
very little extra-spindle actin exists in some cells in which
spindle actin is prominent. Nonetheless, final resolution of
these arguments will only come about when one has quantitative
data describing the distribution of actin-containing filaments
in spindles or when one can otherwise decide whether there are
actin-containing filaments associated with each and every
chromosomal spindle fibre (and/or chromosome). Serial-section
reconstruction of 4 kinetochores in one Haemanthus endosperm
cell at metaphase confirms such an association, but more cells
need to be studied in this regard. Fluorescence microscope
analysis of spindle actin purports to show such a regular
arrangement of actin (a very important statement to be able to
make in stating a case for actin functioning in mitosis), and

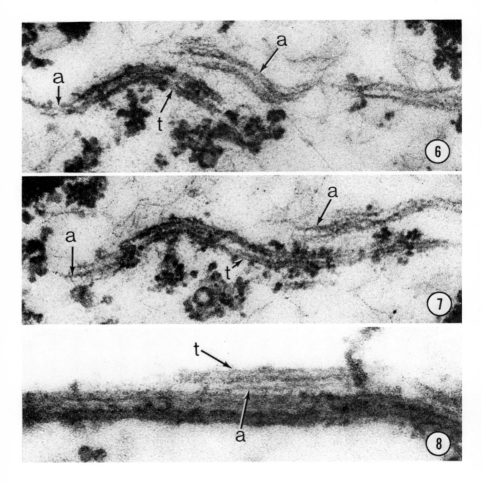

Figs. 6 and 7. Electron micrographs from consecutive
sections of a glycerinated Haemanthus endosperm metaphase cell
that was reacted with rabbit skeletal muscle HMM; these illustrate the closeness of spindle microtubules and actin-containing filaments, as described in the text. Some actin-containing filaments are labelled a, and some microtubules are labelled t. Both are X 70,000. Fig. 8 is an electron micrograph of a section of a glycerinated Haemanthus endosperm cell
that was treated with HMM, illustrating the closeness of an
actin-containing filament (labelled a) and microtubules (one
of which is labelled t). X 115,000.

I discuss these data next.   Unfortunately for proponents of
actin function in mitosis, such as me, these data are not compelling.   I discuss first the data using fluorescent labelling
via antibodies, and then the data using fluorescently labelled
HMM.

C. Studies of Actin (and Myosin) in Spindles using Antibodies
    and Fluorescence Microscopy

     I first outline the general approach, then describe the
individual experiments, and then discuss critically the interpretation of these experiments.
     The general approach in localising intracellular antigens
using fluorescence and antibodies is as follows (see Goldstein, 1976, for example, for references).   One purifies an
antigen, and then obtains antibodies to the antigen by inject-
ing the antigen into an animal (e.g., a rabbit).   Serum is
taken from the animal prior to injections (this is the control
'pre-immune serum') and serum is taken from the animal after
injections (this contains the antibodies).   One determines
that the antibodies are specific for the antigen injected (and
not to minor components in the injected solution) and one determines that the pre-immune serum does not contain antibodies
which react with the cells (e.g., Trenchev & Holborow, 1976).
Then one adds the specific antibodies to a cell which contains
the antigen (the cell must first be made permeable to the
antibodies), and, after rinsing away antibodies which are not
tightly bound, one identifies the position of the antibodies
(and hence the specific antigen) by means of a fluorescent
marker.   In the 'direct fluorescence' technique the fluorescence is inherent in the antibodies themselves.   In the 'indirect fluorescence' technique one adds fluorescently-labelled
antibodies from a second animal which are directed against the
antibodies from the first animal: one injects antigens into
rabbits, for example, reacts rabbit antibodies with the cell,
and then determines the localisation of the rabbit antibodies
by using commercially available fluorescently-labelled goat
antibodies directed against rabbit immunoglobulins.
     Experiments using antibodies against actin and indirect
immunofluorescence of spindles in PtK1 cells in culture were
described by Cande et al. (1977).   Antibodies were prepared
against sodium-dodecyl-sulphate (SDS) - treated calf thymus or

chick fibroblast actin by injection into rabbits. Antibodies
against tubulin were made in a similar way using SDS-treated
porcine brain tubulin. The specificity of the actin antibodies was confirmed by 3 lines of evidence (Lazarides, 1975):
(1) there was a single precipitin line against partially purified actin in immunoelectrophoresis and double immunodiffusion
tests; (2) pre-absorption of the antibody with heat-treated
actin destroyed the fluorescence staining of cells and of myofibrillar I-bands; pre-absorption with native tubulin had no
effect on these staining reactions; and (3) there were different staining patterns when cells were stained with different
antibodies (viz., against actin, tropomyosin, and α - actin-
in). Similar studies demonstrated the specificity of the
antibodies against tubulin.

The techniques used to stain the PtK1 cells were as follows. Mitotic cells were gently lysed in a medium containing
0.04% Triton-X 100 and tubulin, under conditions in which anaphase movements continue for 20 min. or so (Cande et al.,
1974). Then the cells were fixed (in 3.7% formaldehyde, pH
6.4, in a salt solution containing 2% Carbowax), then rinsed
with salt solution, and then incubated in non-immune goat serum; this last step is done in order to absorb goat serum to
all non-specific absorption sites, so that later staining with
fluorescently-labelled goat serum antibodies against rabbit
immunoglobulin will be specific. Then, after rinsing, cells
were reacted with the rabbit antibodies, then rinsed, then
reacted with fluorescently-labelled goat antiserum against
rabbit immunoglobulins (antibodies), and then rinsed, mounted
in glycerol, and observed. Some cells were not lysed, but instead were fixed directly in formaldehyde; then they were
treated with 50% acetone, then 100% acetone, and then 50%
acetone (to make the cells permeable to antisera), and then
treated with non-immune goat serum and processed as were the
others.

The results were as follows, considering first the results for antibodies against actin. Staining with pre-immune
serum gave no fluorescence. Staining non-lysed cells resulted
in spindle staining, but background cytoplasmic fluorescence
tended to mask details. Staining lysed metaphase cells demonstrated an amorphous staining of the spindle pole (a "halo")
and "fibers running from chromosome to pole..." (Cande et al.,
1977). (These cells have 11 chromosomes; the exact number of
chromosome-to-pole fibres could not be counted, for technical

reasons, but 6 such fibres were seen in one cell.) No such
fibres were seen to cross the metaphase plate and, consistent
with the idea that these correspond to chromosomal spindle
fibres, the fibres were of unequal length in prometaphase. In
anaphase the fluorescent fibres were shorter than in metaphase
(as chromosomes moved closer to the poles), there were still
amorphous "halos" at the poles, and no fluorescent fibres were
seen interzonally (though later in anaphase there was diffuse
interzonal fluorescence). Staining with antibodies against
tubulin, using lysed cells, gave different results than did
staining with antibodies against actin: fluorescent fibres
were seen between chromosomes and poles, with no fibres extending across the metaphase plate, but rather than a polar
"halo" there were distinct astral fibres radiating into the
cytoplasm. Also, there were distinct interzonal fibres in
anaphase when antibodies against tubulin were used. When
non-lysed cells were used, the tubulin antibody was localised
more-or-less the same way as in lysed cells, with the differences that there were fluorescent fibres which crossed the
equator, in metaphase, and that the chromosome-to-pole fibres
were somewhat less distinct. From electron microscopic observation of cells treated with antibodies against actin,
Cande et al. (1977) concluded that the spindles were severely
extracted, that the numbers of spindle microtubules were "substantially less..." than in non-treated cells, and that there
was "fuzzy amorphous..." material near microtubules which "may
represent antibody binding to poorly preserved actin filaments..." (Cande et al., 1977). Cande et al. (1977) argue
from these results that, though they can not rule out the
possibility that actin is trapped in the spindle, they have
ruled out other possible artifacts; since the results from
localising actin differ from those from localising tubulin,
they argue, this "rules out the possibility that the actin
localization is just a visualization of the microtubule distribution in the formalin-fixed spindle..." (Cande et al.,
1977). They conclude that "actin antigenicity is associated
with the mammalian mitotic spindle, and that the spatial distribution of this antigenicity changes during spindle formation and function...", and, further, that "actin in a fixable
form is a component of the chromosomal spindle fibers..."
(Cande et al., 1977).
    My criticisms of these interpretations deal with the en-
tire immunofluorescence technique as it has been applied to

date in studying cellular actin, myosin, and tubulin (e.g.,
Weber, 1976; Brinkley et al., 1976; Forer et al., 1976; Franke
et al., 1977; and Pepper & Brinkley, 1977, to cite some recent
examples which used immunofluorescence and antibodies against
tubulin). Thus I will describe the experiments using antibodies against myosin before I discuss those just summarised
dealing with antibodies against actin.

Goode (1975), Mohri et al. (1976), and Fujiwara & Pollard
(1976) have described immunofluorescence of mitotic cells
after reaction with antibodies against myosin. Goode (1975)
used antibodies against chicken myosin and, using the direct
fluorescence technique, added the antibody to glycerinated
chicken embryonic cells in mitosis. He found fluorescent material "occasionally present in or around the mitotic spindle
of myocardial cells..." (Goode, 1975); the fluorescent images
(e.g. Figs. 5, 9, and 10 in Goode, 1975) seem to show fibres
extending between chromosomes and poles. Mohri et al. (1976)
used rabbit antibodies directed against Asterias egg myosin,
in the indirect fluorescence technique, and they stained MA
which were isolated from sea urchins using 1M glycerol (the
method of Sakai et al., 1977). The isolated MA were fixed in
10% formalin, were air dried on a slide, were rinsed, were incubated in rabbit antibodies, were rinsed, were incubated in
fluorescent goat antibodies (against rabbit immunoglobulins),
were rinsed, were placed in glycerol, and were studied. "Non-
immune..." rabbit serum gave no reaction. Antibodies against
myosin stained the MA slightly to give diffuse staining
throughout, except for chromosomes, which did not stain, and
except for discrete and brighter staining of spindle poles,
which appeared as amorphous "halos".

Fujiwara & Pollard (1976) used direct immunofluorescence
to stain mitotic spindles in HeLa cells in culture, using
antibodies against myosin, and myosin prepared from human
blood platelets. The specificity of the antibodies was tested
carefully using various methods. For example, using immunoelectrophoresis, the antibodies did not react with actin, or
with the head (ATPase) region of myosin, but did form single
precipitin lines with purified platelet myosin or with purified rod portion of platelet myosin (the non-ATPase portion),
and they did form a single precipitin line when reacted with
crude platelet extract. (Crude platelet extract and crude He-
La cell extract consisted of the material which was soluble
when whole human platelets, or HeLa cells, were broken open

into 40 mM sodium pyrophosphate solution.) Further, using double immunodiffusion, single (confluent) precipitin lines were formed with purified platelet myosin, rod portion of platelet myosin, crude HeLa cell extracts, and crude platelet extracts, whilst there was no reaction with myosin from human skeletal or cardiac muscle. Finally, using tandem crossed immunoelectrophoresis, there was one reacting component in crude platelet extract which was identical in position with (i.e., the line fused with that of) purified myosin. (It might be relevant to note that these tests were done with antibodies which were not fluorescently labelled.) Further care was taken to ensure that fluorescently-labelled antibodies did not react non-specifically: only those with 2-5 dye molecules per protein molecule were used for cytochemical localisations. In addition, specific antibodies against myosin were purified (by affinity column chromatography) from the fluorescently-labelled immunoglobulins.

The staining procedures were as follows. HeLa cells were treated with acetone at -20°C, for 5 min., and then air-dried. Air-dried cells were treated with antibodies, rinsed, and then placed in 90% glycerol and studied. Equivalent results were obtained when acetone-fixed cells were post-fixed with 1 to 1.5% formalin. Various controls were run, to assure that staining was due to specific antibodies, and these included staining with pre-immune serum and staining with antibodies pre-absorbed with myosin, neither of which gave staining of spindles. A further control was the absence of staining of spindles when non-human cell lines were studied, namely in rat-kangaroo (PtK2) cells or in salamander cells.

Fujiwara & Pollard (1976) found staining (fluorescence) throughout the HeLa cell cytoplasm, but there were higher intensities of staining in the spindle and in the cleavage furrow than elsewhere. Spindle staining was between the chromosomes and poles, in both metaphase and anaphase, but this was diffuse, and was not seen as discrete fibres. In anaphase the interzone also stained, without visible fibres, "but much of this intensity is attributable to the cleavage furrow surrounding this part of the cell..." (Fujiwara & Pollard, 1976). Purified antibody against myosin stained the spindle less intensely than did whole immunoglobulins; Fujiwara and Pollard discussed possible explanations for this, but because of this result they conclude, cautiously, that they "believe that myosin is concentrated in the mitotic spindle, but feel that this point needs further investigation..." (Fujiwara & Pollard, 1976).

In evaluating the immunofluorescence results, both for actin and myosin distributions, I think three important criteria need be kept in mind: (1) are the antibodies specific for the antigen in question? (2) If the antibodies are specific for that antigen (i.e. if they do not react with minor components which may have been injected), do any other proteins in the cell share antigenic determinants with that antigen? (In this case antibodies would react both with the antigen and with the other components.) (3) Is there movement of antigen during the procedures of fixation and antibody treatment? I will look at these criteria in turn.

The various experiments using fluorescently marked antibodies generally have included sufficient controls to demonstrate that the antibodies are specific for the particular antigen in question, though some caution should be exerted even here because Fujiwara & Pollard (1976) demonstrate that non-specific antibodies (which react with more than one component in the antigen preparations) can pass some of the usual tests for specificity.

The second criterion is one of the main reasons why the results can not be accepted as demonstrating spindle actin and spindle myosin (or, for that matter, as demonstrating spindle tubulin by means of fluorescence microscopy): no-one has demonstrated that there are no other components in the cell which share antigenic determinants with the antigen in question. To illustrate what I mean by this, one can look at the experiments done by Fujiwara & Pollard (1976), who demonstrated that when antibodies against myosin reacted with crude cell extract, only one component in the extract reacted with the antibodies. (The antibodies against tubulin and against actin have not been tested even in this way, as far as I am aware.) This commendable experiment is closer than most to a rigorous control, but two things are missing. First, while the antibody against myosin reacts with only one soluble cell component, there may be several components in the insoluble (discarded) fraction with which the antibody reacts, and which would show up as fluorescence in the cytochemical studies. Second, and perhaps more importantly, this test, that the antibody reacts only with one component in the cell extract, was done under conditions that were quite different from the cytochemistry: the immunoelectrophoresis was done with more-

or-less native proteins while the cytochemistry was done on proteins that were treated with cold, 100% acetone and then air dried. (An additional difference in this particular case is that the immunoelectrophoresis was done using not-fluorescently labelled antibodies, while the cytochemistry was done using fluorescently-labelled antibodies.) Thus, while no native proteins might share antigenic determinants with the antigen in question, denatured proteins might do so. Hence, testing for other reacting components in the cell needs to be done under the same conditions that the cells are treated for immunofluorescence.

This point is important, and not just nitpicking, for there are data which show clearly that denaturation can expose antigenic determinants that otherwise would not react with the antibodies. As one example, one could cite the data of Mabuchi & Okuno (1977): antibody against Asterias egg myosin did not react with native (not-denatured) platelet myosin, or with native rabbit skeletal muscle myosin, but did react with these other proteins after they were treated with 0.05% SDS. Hence the treatment with SDS (denaturation) exposed antigenic determinants that were recognized by the antibodies, but that were otherwise hidden in native proteins. As another example, perhaps even more relevant to the cases under discussion, one could cite the data of Nishino & Watanabe (1977a,b): antiserum directed against 14S dynein from Tetrahymena does not react with native tubulin or native actin, but does react with both denatured tubulin and denatured actin (though not with denatured myosin, tropomyosin, or troponin). Hence there are antigenic determinants in common between actin, tubulin, and dynein when these proteins are denatured. Thus it is quite possible that antibodies thought to be specific for actin, for example, would react with denatured dynein and denatured tubulin, and that the intracellular localisations observed would be combinations of the positions of all 3 proteins, and perhaps others. As a final example, one can consider the γ - subunit of glycogen phosphorylase kinase in the Pacific dogfish: this subunit has an amino acid composition identical to that of actin, it reacts with HMM to polymerise into arrowhead-decorated filaments, and it stimulates the Mg^{++}-ATPase of HMM (Fischer et al., 1975, and E.H. Fischer, personal communication; I am indebted to Sally Tobin for pointing this work out to me). The γ - subunit is ordinarily bound tightly into the glycogen phosphorylase kinase enzyme, under which condit-

tion it does not react with HMM (E.H. Fischer, personal communication), but under denaturing conditions (such as those
used for the cytochemistry) the γ - subunit might be freed,
and might be able to react with antibodies against actin.
Hence if one is to accept the cytochemical localisations, one
must rigorously test whether the antibodies in question will
react with more than one protein in the cell, and this test
must be done under the same conditions as the cytochemistry is
done. Otherwise one can question the localisations determined
cytochemically as being due to several proteins, and not just
to the antigen.

It might be relevant to point out a straightforward way
in which this control can be run. Various people have demonstrated that antibodies still react with SDS-treated antigens
(e.g., Stumph et al., 1974; Mabuchi & Okuno, 1977; and Olden &
Yamada, 1977). Thus one can perform electrophoresis of whole
cells in weak SDS: some components will remain at the origin,
and others will migrate. Then one gel can be stained, to see
where the bands are, whilst a parallel gel can be fixed and
treated as if it were the cell being studied (e.g., with acetone, and air-dried; or with formaldehyde, followed by acetone, etc.); then the treated gel can be reacted with fluorescently labelled antibody, to see whether the antibody will
bind solely to the antigen or will bind to other bands as
well. This, then, performs the required test of specificity
under the exact conditions of the cytochemistry. (A somewhat
different way has been used: Olden & Yamada (1977) reacted
such gels with non-labelled antibodies and then by staining
and microdensitometry measured increased protein concentrations at the bands in question.) Only by doing some experiment of this kind will one be able to have confidence in the
fluorescence microscopy results, and to believe that the antibodies bind solely to the antigen in question.

The third criterion, adequacy of fixation in immobilising
the antigen, should also be considered: 'inadequate fixation'
is another serious criticism which can be levelled against the
fluorescence microscopy results. Two lines of evidence demonstrate that the fixation using formalin and acetone (or variants of this) is inadequate, in that it permits relocation of
antigen. First is electron microscopic study of cells treated
in this way. There is agreement amongst several groups of
workers that, as seen in sectioned material, the fixation procedures commonly used for fluorescence microscopy studies

preserve only some microtubules (e.g. Cande et al., 1977; Sato et al., 1976), or no microtubules (Forer et al., 1976; or as illustrated in Figs. 1-3, and 6-11 in Pepper & Brinkley,1977). Hence, since at least some of the tubulin has moved during this fixation, other antigens also might move. Indeed, Forer et al. (1976) argue that using fluorescence microscopy one observes only tubulin which binds to another component when the microtubules break down, and hence one sees in fluorescence microscopy the position of the other component and not necessarily the original position of the tubulin. The second kind of evidence indicating translocation of antigen is the appearance after fixation: cells fixed for immunofluorescence look quite different from living cells as viewed in phase-contrast microscopy. The differences are striking: in living cells, chromosomes are phase dark and spindles are light against the darker cytoplasmic background and are clearly separated from the cytoplasm, whereas in cells fixed for fluorescent antibody studies chromosomes may be phase bright, or not seen, and spindles are not seen or are only weakly set-off from the cytoplasm. [In making this evaluation, one needs to use spindles treated with pre-immune serum or the like, in which there is no specific labelling, because addition of antibody to only one part of the cell will change the concentrations of protein in that part of the cell. Hence, I considered Fig. 19 and 20 of Fujiwara & Pollard (1976), and compared it with living cells illustrated in Robbins (1961) and Robbins & Gonatas (1964). Or I considered Figs. 3-5 in Cande et al., 1977 and compared these with living cells illustrated in Roos (1976) and Brenner et al. (1977). Or I considered Figs. 12 and 13 in Franke et al. (1977) compared with living cells illustrated in Figs. 1 and 2 of Franke et al. (1977).] Since contrast in phase-contrast microscopy is a measure of relative path difference, which in turn is a measure of path length and concentration of dry matter, these alterations mean that relative concentrations have changed and that some intracellular materials have relocated; hence there is the clear possibility that antigens have moved from their original positions. This seems not to be due to the glycerol in which the final products are mounted, because cells fixed and prepared for electron microscopy and looked at after they are impregnated (and mounted) in Epon or Araldite still look more-or-less like cells in vivo (see, e.g., pictures of such cells in Bajer and Molé-Bajer, 1969), though, of course, one has here the complication that a colouring agent (osmium) has been added.

It might be relevant to consider controls that might help decide if relocation does indeed occur during fixation. One way might be to use fluorescently labelled antigen that is added with the fixative (and/or later steps) to see if the antigen selectively binds to some structure, and/or if the antigen relocates during subsequent procedures.

Experiments by Goldman et al. (1977) are relevant in discussing the reliability of fluorescence microscopy using antibodies; these experiments compared electron microscopic identification of bundles of actin-containing filaments ("stress fibres") with immunofluorescence localisations using antibodies against actin. Goldman et al. (1977) found clear discrepancies, in that "stress fibres" were present in some cells, in normal numbers, as identified electron microscopically, yet were not seen using fluorescence microscopy. Hence, for whatever reason, one can not necessarily rely on fluorescence microscopy alone (see Goldman et al., 1977, for discussion of this).

I have raised objection to the cytochemical localisations of actin and myosin, because there may be other cellular proteins with similar antigenic determinants. A counter-argument is implicit in the discussion by Cande et al. (1977) that, as determined by immunofluorescence, the actin localisation is different from the tubulin localisation, and this "rules out the possibility that the actin localization is just a visualization of the microtubule distribution..." (Cande et al., 1977). This argument is not compelling. For one, we know that kinetochore-associated microtubules respond differently from other spindle microtubules when cells are treated in various ways (e.g., Lambert & Bajer, 1977), and that astral fibres respond yet differently (e.g., Forer et al., 1976). Whatever the reason is, there are some differences between the different kinds of spindle fibres, and these differences could give rise to the different staining and yet still not be related to differences in actin or tubulin distribution. Until one can rule out that the antibodies react with several proteins under the conditions of the cytochemistry, one can not rule out alternative interpretations of the data.

It is also relevant to point out that even were all my objections overruled, one still needs definitive proof that each and every chromosomal spindle fibre has actin. This was suggested but not definitively proven in the cases cited

above, because there were too many chromosomes per cell. Such
proof can probably only be done by studying some cell with
only a few spindle fibres, such as crane fly spermatocytes,
that have only 5 chromosomal spindle fibres per spindle (e.g.,
Forer, 1976).

In summary, I have concluded that the immunofluorescence
localisation of actin and myosin in spindles is not compelling, because there is the possibility (or even likelihood)
that the antibodies reacted with several proteins under conditions of the cytochemical treatment, and because there is reason to believe that antigens are not immobilised by the fixation procedure.

I now consider experiments using fluorescently labelled
heavy meromyosin (HMM).

D. Studies of Actin in Spindles Using Fluorescently Labelled
    HMM

I first outline the general approach, then describe the
individual experiments, and then discuss the interpretation of
these experiments.

The general approach is as follows. HMM binds specifically to actin to form arrowheads, and this binding is blocked
by ATP or pyrophosphate. The general idea, then, is to make
the HMM fluorescent (by labelling it with a fluorescent marker), and then to visualize, using fluorescence microscopy, the
localisation of the HMM after the HMM has reacted with the
actin in the cell (Aronson, 1965).

Spindle actin has been studied, using fluorescent-HMM, by
Aronson (1965), Sanger (1975a,b), Sanger & Sanger (1976), and
Schloss et al. (1977). Aronson (1965) labelled HMM with
fluorescein; this destroyed 90% of the HMM ATPase, and a large
fraction of the HMM did not bind to myofibrillar actin-containing filaments. When fluorescent-HMM was reacted with myofibrils the fluorescent staining pattern was nonetheless
specific for actin-containing filaments, as demonstrated by
the following controls. The myofibrils stained heaviest in
the I-band, weaker in the A-band (overlap region), and not at
all in the H-gap. Pre-treatment with unlabelled HMM abolished
"all but a very faint trace..." of fluorescent staining, and
addition of pyrophosphate caused loss of most of the staining. Addition of 0.6M KI (which dissolves out the actin-con-

taining filaments) abolished the staining. Finally addition
of fluorescein itself to untreated, glycerinated myofibrils
resulted in a staining pattern quite different from that using
fluorescent-HMM. Aronson (1965) added fluorescent-HMM to isolated sea urchin zygote spindles, and found that both the
spindles and the chromosomes bound the HMM, "although in this
instance no attempt was made to establish the specificity of
the binding..." (Aronson, 1965), by means of pyrophosphate
addition or the like.

Sanger (1975a,b) and Sanger & Sanger (1976) studied fluorescent-HMM labelling of PtK2 spindles, and established the
specificity of their fluorescent-HMM by reacting the HMM with
myofibrils, and by using pyrophosphate or pre-treatment with
unlabelled HMM to block the reaction. Sanger (1975b) glycerinated rat kangaroo (PtK2) cells in the cold, using 50% glycerol; then he reacted them with fluorescent-HMM (in 25% glycerol), rinsed them, and placed them in 50% glycerol for view-
ing. In metaphase cells he saw discrete fluorescent fibres,
6-9 per cell (though it was difficult to determine the exact
number), extending between chromosomes and poles, and the
chromosomes were unstained except for a small "dot" where the
fibres ended (the kinetochores). There were no continuous
fibres, and there were no astral fibres, though there were
round spots at the poles. In anaphase the chromosome-to-pole
fibres were shorter, and there were no interzonal fibres,
though diffuse fluorescence appeared interzonally at late anaphase. Absence of the fluorescence after treatment of mitotic
cells with ATP or pyrophosphate was not reported in Sanger
(1975b); Sanger & Sanger (1976) did report this control,
though, namely that 10mM ATP blocked the staining of the
chromosome-to-pole fibres, though some fluorescence remained
near the poles. To quote Sanger (1975b), "we conclude...that
actin is present in...chromosomal spindle fibers of rat kangaroo cells."

Schloss et al. (1977) labelled S1 with fluorescein, and
then separated the fluorescent-S1 (using DEAE cellulose col-
umns) into portions with varying amounts of fluorescein per
protein, namely, 1,2,2.5, or 4-5 fluorescein molecules per
protein molecule. This was necessary in order to avoid non-
specific staining (which was not reversible using pyrophosphate) which occurred when there were 4-5 fluorescein molecules per S1 molecule. The specificity of the fluorescent-S1
was tested in a number of ways. The S1 gave arrowheads with

muscle actin, and the arrowheads were removed by pyrophosphate. The S1 maintained 68% of its actin-activated ATPase
after labelling with fluorescein. Stress fibres were stained
with fluorescent-S1, and this staining was blocked completely
with 4mM pyrophosphate or by pre-treatment with unlabelled S1,
whereas pre-treatment with bovine serum albumen had no effect.
In labelling spindles, PtK1 cells were glycerinated for as
little as 2-4 minutes prior to treatment with fluorescent-S1.
In metaphase cells, chromosome-to-pole fluorescence was seen,
as discrete fibres. In anaphase these fibres shortened, and
there was no interzonal fluorescence. In both stages, the
cell cortex also was fluorescent. Both 4mM pyrophosphate and
pre-treatment with unlabelled S1 blocked the fluorescence
staining, except for staining of the poles. Electron microscopic observations were made on these cells, as described
above, and actin-containing filaments were seen. The authors
conclude that actin is in the spindle and that "an actomyosin
like contractile system may play a role in chromosome movement
during mitosis and meiosis..." (Schloss et al., 1977). I now
discuss the results of fluorescent-HMM labelling of spindles.
        Several points need be made in assessing these results.
First is the possibility that when pyrophosphate (or ATP)
blocks the staining, this may be due to dissolving out a non-
specific binding component rather than blocking the reaction;
mM ATP, for example, solubilizes several muscle proteins
(e.g., Cooke, 1972). This criticism is especially relevant to
the results of Sanger (1975b) and Sanger & Sanger (1976),
since treatment with 10mM ATP is the only control they report
for staining spindles with fluorescent HMM. A control for
this is first to block (or remove) the staining by treatment
with ATP or with pyrophosphate, and then to show that the
staining is regained after rinsing and subsequent treatment
with fluorescent-HMM (or fluorescent-S1). Schloss et al.
(1977) did exactly this control for staining stress fibres
with fluorescent-S1, but it is unclear from their statement of
the controls for staining spindles (p.800 of Schloss et al.,
1977) whether they did this control for staining spindles.
Hence it is not clear whether the controls which have been run
to date satisfactorily demonstrate specificity. Second is a
major problem with the technique as used by all groups: HMM
binds to actin, specifically, to give arrowheads, but HMM
might bind to other cell components, to not give arrowheads,
and to not give binding visible electron microscopically, yet

such binding would be visible using fluorescence microscopy.
Examples of components HMM might bind to include ribosomes and
polyribosomes: ribosomes are present in spindles (e.g., Hartman & Zimmerman, 1968; Goldman & Rebhun, 1969), as are polyribosomes (e.g., Forer & Jackson, in preparation), and both
ribosomes and polyribosomes will bind to myosin and coprecipi-
tate with myosin (e.g., Heywood et al., 1968). Thus, the
fluorescent-HMM might bind to spindle ribosomes which are oriented between chromosomes and poles. I do not know if this
binding would be sensitive to pyrophosphate, but it might be,
since it is sensitive to KCl in exactly the same manner as
actin and myosin (Heywood et al., 1968). Another example might
be glycogen phosphorylase kinase (Fischer et al., 1975), as
discussed previously when considering antibodies against
actin. Thus other controls are needed to verify that the
binding is indeed to actin and only to actin. One such control might be to block the binding by means of DNAase treatment, or thrombin treatment, which would break down actin-containing filaments (review in Forer, 1978) but which would not
be expected to alter ribosomes or polyribosomes. Or one might
run electrophoretic gels of the entire cell and then, as in
the control proposed for staining by actin antibodies, one
could determine if the HMM (or S1) bound to only the actin
band or to other bands as well.

A third point is that, until one studies cells in which
there are only a few spindle fibres, one does not have compelling evidence to prove that each and every spindle fibre
has actin, as discussed previously. The final point is one
that has been discussed, and that is the problem of relocation
during glycerination. Schloss et al. (1977) shortened the
glycerination time to 2-4 minutes to minimize this problem,
and they also reported that formalin-fixed cells (not glycerinated) have staining patterns identical to those of glycerinated cells. However, relocation can occur during formalin
fixation too, as argued above. Perhaps the best argument
against this would be to use intranuclear spindles, or to demonstrate unequivocally that each and every spindle fibre always contains actin.

In summary, there are qualifications to the fluorescent-
HMM evidences that actin is in spindles, in that proper ATP or
pyrophosphate reversibility controls have not been run, in
that non-actin components in the spindle might bind HMM (reference was made to myosin-ribosome binding), and in that it

still has not been demonstrated that each and every chromosomal spindle fibre contains actin (or HMM-binding material);
demonstration that each spindle fibre always contains actin
would argue forcibly against the "relocation" problem.

Two additional points need to be considered. First is
the "contradiction" between the electron microscopic localisation of actin and the fluorescence microscopic localisation of
actin. This 'discrepancy', discussed by Schroeder (1976),
Cande et al. (1977) and Schloss et al. (1977) is essentially
that actin-containing filaments are prominent interzonally, as
studied electron microscopically, yet little fluorescence
staining is seen interzonally, using either antibodies against
actin or fluorescent-HMM. To me this 'discrepancy' is difficult to discuss because of the severe criticisms I have of the
fluorescence studies, that, because of missing controls, it is
possible that much of the localisation is due to binding with
components other than actin. Even disregarding my objections
to the fluorescence data, it seems clear to me that one does
not really even know if a 'discrepancy' exists, for neither
the electron microscopic distribution of filaments nor the
fluorescence distribution has been presented in a quantitative
manner. Until one knows even roughly how many microfilaments
there are in the various regions, and compares these data with
even rough amounts of fluorescence, one does not know if a
'discrepancy' exists, let alone what to make of it.

The second point is the striking similarity in results
using antibodies against actin and using fluorescent-HMM: both
stain discrete fibres between chromosomes and poles, both
stain weakly interzonally, both do not stain astral fibres,
and both stain the spindle poles as diffuse circular areas.
And this pattern of staining is quite different from that seen
after staining with antibodies against tubulin. Even though
there are severe criticisms of each technique, separately, one
nonetheless could argue that the staining pattern indeed re-
flects spindle actin, because the two different techniques
give the same result, and this result is different from results of staining using other fluorescent antibodies. This
argument has some force, but because of the uncertainties discussed above one can still think of simple ways to explain the
congruence. For example, that actin antibodies bind to non-
actin components would mean that these non-actin components
had antigenic determinants in common with (or similar to)
actin. If these antigenic determinants are chemically similar

to the HMM-binding site of actin, then both the antibodies and
the HMM would be expected to bind to the same components.
That is to say, if there is any binding to non-actin spindle
components (and I have argued that the controls have not at
all ruled this out), then one might expect both antibodies and
HMM to bind to the same components, since these non-actin components would be similar to actin. Thus while it is tempting
to consider that the 2 sets of data together indicate that
spindle fibres indeed contain actin, until more controls are
done these results are not unequivocal.

In conclusion, the chemical nature of spindles is largely
unknown, but microtubules are universally associated with
chromosomal spindle fibres; these fibres are the most likely
sites of force production. The microtubules occupy only a
small fraction of the spindle, and it is not known whether it
is the microtubules or other spindle fibre components that
produce the force for chromosome movement. To me, the most
likely hypothesis is that actin is involved in producing the
force. I have argued that actin is indeed a bona fide spindle
component, and I have summarised the various evidences on this
point. I have summarised the counter-arguments, and dismissed
many of them. But some counter-arguments have not been eliminated. One question to be answered in determining whether
actin functions in chromosome movement is to see if actin is
associated with each and every chromosomal spindle fibre.
Quantitative electron microscopy can be done to demonstrate
this (or deny this), but so far has not been done in sufficient detail. Fluorescent-HMM data (and fluorescence microscope data localising actin antibodies) suggest that chromosomal spindle fibres contain actin, but these experiments are
not definitive because further controls need be done.

ACKNOWLEDGMENTS

I gratefully acknowledge the help of several colleagues:
Dawn Larson has helped with immunological references, as have
Saul and Elena Puszkin; Sally Tobin told me of her interesting
experiments (with C. Laird), and of the work of Edmond Fischer; Edmond Fischer told me details of his work; Brent Heath
helped with critical comments; and Bill Jackson helped with
critical scientific comments, and, as well, grammatical ones
that have helped ease some ponderous sentences. I appreciate

the excellent secretarial help of Dorothy Gunning. My work was supported by grants from the National Research Council of Canada.

REFERENCES

Allen, R.D., and Taylor, D.L. (1975). In "Molecules and Cell Movement" (S. Inoué and R.E. Stephens, eds.), pp. 239-258. Raven Press, New York.

Aronson, J.F. (1965). J. Cell Biol. 26, 293-298.

Bajer, A.S. (1973). Cytobios 8, 139-160.

Bajer, A., and Mole-Bajer, J. (1969). Chromosoma 27, 448-484.

Bajer, A.S., Mole-Bajer, J., and Lambert, A.-M. (1975). In "Microtubules and Microtubule Inhibitors" (M. Borgers and M. de Brabander, eds.), pp. 393-423. North-Holland Publishing, Amsterdam.

Behnke, O., Forer, A., and Emmersen, J. (1971). Nature, Lond. 234, 408-410.

Bibring, T., and Baxandall, J. (1971). J. Cell Biol. 48, 324-339.

Bray, D., and Thomas, C. (1976). J. molec. Biol. 105, 527-544.

Brenner, S., Branch, A., Meredith, S., and Berns, M.W. (1977). J. Cell Biol. 72, 368-379.

Brinkley, B.R., Fuller, G.M., and Highfield, D.P. (1976). In "Cell Motility. Book A: Motility, Muscle and Non-muscle Cells" (R. Goldman, T. Pollard and J. Rosenbaum, eds.), pp. 435-456. Cold Spring Harbor Laboratory, Cold Spring Harbor, N.Y.

Bruggmann, S., and Jenny, E. (1975). Biochim. Biophys. Acta 412, 39-50.

Cande, W.Z., Lazarides, E., and McIntosh, J.R. (1977). J. Cell Biol. 72, 552-567.

Cande, W.Z., Snyder, J., Smith, D., Summers, K. and McIntosh, J.R. (1974). Proc. natn. Acad. Sci. U.S.A. 71, 1559-1563.

Cohen, W.D., and Rebhun, L.I. (1970). J. Cell Sci. 6, 159-176.

Cooke, R. (1972). Biochem. biophys. Res. Comm. 49, 1021-1028.

Cornman, I. (1944). Am. Nat. 78, 410-422.

Dietz, R. (1969). Naturwissenschaften 56, 237-248.

Dietz, R. (1972). In "Chromosomes Today Vol. 3" (C.D. Darlington and K.R. Lewis, eds.), pp. 70-85. Longman, London.

Euteneuer, U., Bereiter-Hahn, J., and Schliwa, M. (1977). Cytobiologie 15, 169-173.

Fischer, E.H., Becker, J.-U., Blum, H.E., Byers, B., Heizmann, C., Kerrick, G.W., Lehky, P., Malencik, D.A., and Pocinwong, C. (1975). In "Molecular Basis of Motility: Colloqium-Mosbach 1975" (L. Heilmeyer, J.C. Rüegg, and Th. Weiland, eds.), pp. 137-158. Springer Verlag, Berlin-Heidelberg-New York.

Forer, A. (1966). Chromosoma 19, 44-98.

Forer, A. (1969). In "Handbook of Molecular Cytology" (A. Lima-de-Faria, ed.), pp. 553-601. North-Holland Publishing Company, Amsterdam and London.

Forer, A. (1974). In "Cell Cycle Controls" (G.M. Padilla, I. L. Cameron, and A.M. Zimmerman, eds.), pp. 319-336. Academic Press, New York and London.

Forer, A. (1976). In "Cell Motility, Book C. Microtubules and Related Proteins" (R. Goldman, T. Pollard, and J. Rosenbaum, eds.), pp. 1273-1293. Cold Spring Harbor Laboratory, Cold Spring Harbor, N.Y.

Forer, A. (1978). In "Principles and Techniques of Electron Microscopy:Vol. 9, Biological Applications" (M.A. Hayat, ed.), in press. Van Nostrand-Reinhold, New York.

Forer, A., and Behnke, O. (1972a). J. Cell Sci. 11, 491-519.

Forer, A., and Behnke, O. (1972b). Chromosoma 39, 145-173.

Forer, A., and Behnke, O. (1972c). Chromosoma 39, 175-190.

Forer, A., and Blecher, S.R. (1975). J. Cell Sci. 19, 576-605.

Forer, A., and Brinkley, B.R. (1977). Can. J. Genet. Cytol. 19, 503-519.

Forer, A., and Goldman, R.D. (1972). J. Cell Sci. 10, 387-418.

Forer, A., and Jackson, Wm. T. (1975). Cytobiologie 10, 217-226.

Forer, A., and Jackson, Wm. T. (1976). Cytobiologie 12, 199-214.

Forer, A., Kalnins, V., and Zimmerman, A.M. (1976). J. Cell Sci. 22, 115-131.

Franke, W.W., Seib, E., Osborn, M., Weber, K., Herth, W., and Falk, H. (1977). Cytobiologie 15, 24-48.

Fujiwara, K., and Pollard, T.D. (1976). J.Cell Biol. 71,

848-875.

Fuseler, J.W. (1975). J. Cell Biol. 64, 159-171.

Gawadi, N. (1971). Nature, Lond. 234, 410.

Gawadi, N. (1974). Cytobios 10, 17-35.

Gibbons, B.H., Ogawa, K., and Gibbons, I.R. (1976). J. Cell Biol. 71, 823-831.

Gibbons, I.R., Fronk, E., Gibbons, B.H., and Ogawa, K. (1976). In "Cell Motility. Book C: Microtubules and Related Proteins" (R. Goldman, T. Pollard and J. Rosenbaum, eds.), pp. 915-932. Cold Spring Harbor Laboratory, Cold Spring Harbor, N.Y.

Goldman, R.D., and Rebhun, L.E. (1969). J. Cell Sci. 4, 179-209.

Goldman, R.D., Schloss, J.A., and Starger, J.M. (1976). In "Cell Motility. Book A: Motility, Muscle and Non-muscle Cells" (R. Goldman, T. Pollard and J. Rosenbaum, eds.) pp. 217-245. Cold Spring Harbor Laboratory, Cold Spring Harbor, N.Y.

Goldman, R.D., Yerna, M.-J., and Schloss, J.A. (1977). J. Supramol. Struct. 5, 155-183.

Goldstein, G. (1976). In "Cell Motility. Book A: Motility, Muscle and Non-muscle Cells" (R. Goldman, T. Pollard and J. Rosenbaum, eds.) pp. 333-336. Cold Spring Harbor Laboratory, Cold Spring Harbor, N.Y.

Goode, D. (1975). Cytobiologie 11, 203-229.

Gröschel-Stewart, U., Ceurremans, S., Lehr, I., Mahlmeister, C., and Paar, E. (1977). Histochemistry 50, 271-279.

Gruzdev, A.D. (1972). Tsitologiya 14, 141-149. (In Russian. English translation as NRC Technical Translation 1758, National Research Council of Canada, Ottawa.)

Hartmann, J.F., and Zimmerman, A.M. (1968). Expl Cell Res. 50, 403-417.

Heath, I.B. (1974). J. Cell Biol. 60, 204-220.

Hepler, P.K., and Palevitz, B.A. (1974). Ann. Rev. Plant Physiol. 25, 309-362.

Heywood, S.M., Dowben, R.M., and Rich, A. (1968). Biochemistry, N.Y. 7, 3289-3296.

Hinkley, R., and Telser, A. (1974). Expl Cell Res. 86, 161-164.

Holtzer, H., Sanger, J.W., Ishikawa, H., and Strahs, K. (1973). Cold Spring Harb. Symp. Quant. Biol. 37, 549-566.

Huxley, H.E. (1972). In "The Structure and Function of

Muscle, Vol. 1, Pt.1" (G.H. Bourne, ed.), pp. 301-387. Academic Press, New York and London.

Huxley, H.E. (1973). Nature 243, 445-449.

Huxley, H.E. (1976). In "Cell Motility. Book A: Motility, Muscle and Non-muscle Cells" (R. Goldman, T. Pollard and J. Rosenbaum, eds.), pp. 115-126. Cold Spring Harbor Laboratory, Cold Spring Harbor, N.Y.

Inoué, S. (1953). Chromosoma 5, 487-500.

Inoué, S. (1964). In "Primitive Motile Systems in Cell Biology" (R.D. Allen and N. Kamiya, eds.), pp. 549-598. Academic Press, New York.

Inoué, S. (1976). In "Cell Motility, Book C: Microtubules and Related Proteins" (R.D. Goldman, T. Pollard, and J. Rosenbaum, eds.), pp.1317-1328. Cold Spring Harbor Laboratory Press, Cold Spring Harbor, N.Y.

Inoué, S., and Ritter, Jr., H. (1975). In "Molecules and Cell Movement" (S. Inoué and R.E. Stephens, eds.), pp. 3-29. Raven Press, New York.

Inoué, S., and Sato, H. (1967). J. gen. Physiol. 50 (No.6, Pt.2), 259-292.

Inoué, S., Fuseler, J., Salmon, E.D., and Ellis, G.W. (1975). Biophys. J. 15, 725-744.

Jacquez, J.A., and Biesele, J.J. (1954). Expl Cell Res. 6, 17-29.

Kersey, Y.F., Hepler, P.K., Palevitz, B.A., and Wessells, N.K. (1976). Proc. natn. Acad. Sci. U.S.A. 73, 165-167.

Kessler, D., Nachmias, V.T., and Loewy, A.G. (1976). J. Cell Biol. 69, 393-406.

Kiehart, D.P., Inoué, S., and Mabuchi, I. (1977). J.Cell Biol. 75, 258a.

LaFountain, Jr., J.R. (1974). J. Cell Biol. 60, 784-789.

LaFountain, Jr., J.R. (1975). BioSystems 7, 363-369.

Lambert, A.-M., and Bajer, A.S. (1977). Cytobiologie 15, 1-23.

Lazarides, E. (1975). J. Histochem. Cytochem. 23, 507-528.

Lazarides, E., and Lindberg, U. (1974). Proc. natn. Acad. Sci. U.S.A. 71, 4742-4746.

Lewis, L.M., Witkus, E.R., and Vernon, G.M. (1976). Protoplasma 89, 203-219.

Mabuchi, I. (1973). J. Cell Biol. 59, 542-547.

Mabuchi, I. (1974). J. Biochem. 76, 47-55.

Mabuchi, I. (1976). J. molec. Biol. 100, 569-582.

Mabuchi, I., and Okuno, M. (1977). J. Cell Biol. 74,251-263.

Mazia, D. (1961). In "The Cell, Vol. 3" (J. Brachet and A.E. Mirsky, eds.), pp. 77-412. Academic Press, New York and London.

McDonald, K., Pickett-Heaps, J.D., McIntosh, J.R., and Tippit, D.H. (1977). J. Cell Biol. 74, 377-388.

McIntosh, J.R., and Landis, S. (1971). J. Cell Biol. 49, 468-497.

McIntosh, J.R., Hepler, P.K., and Van Wie, D.G. (1969). Nature (London) 224, 659-663.

McIntosh, J.R., Cande, W.Z., and Snyder, J.A. (1975). In "Molecules and Cell Movement" (S. Inoué and R.E. Stephens, eds.), pp. 31-76. Raven Press, New York.

McIntosh, J.R., Cande, W.Z., Lazarides, E., McDonald, K., and Snyder, J.A. (1976). In "Cell Motility. Book C: Microtubules and Related Proteins" (R. Goldman, T. Pollard and J. Rosenbaum, eds.), pp. 1261-1272. Cold Spring Harbor Laboratory, Cold Spring Harbor, N.Y.

Mitchison, J.M., and Swann, M.M. (1953). Q. Jl Microsc. Sci. 94, 381-389.

Mohri, H., Mohri, T., Mabuchi, I., Yazaki, I., Sakai, H., and Ogawa, K. (1976). Develop., Growth & Differ. 18, 391-398.

Moore, P.B., Huxley, H.E., and De Rosier, D.J. (1970). J. molec. Biol. 50, 279-295.

Mooseker, M.S. (1976). In "Cell Motility. Book B: Actin, Myosin and Associated Proteins" (R. Goldman, T. Pollard and J. Rosenbaum, eds.), pp. 631-650. Cold Spring Harbor Laboratory, Cold Spring Harbor, N.Y.

Müller, W. (1972). Chromosoma 38, 139-172.

Nachmias, V.T., and Kessler, D. (1976). Immunology 30, 419-424.

Nicklas, R.B. (1971). In "Advances in Cell Biology II" (D.M. Prescott, L. Goldstein and E.H. McConkey, eds.), pp. 225-297. Appleton Century Croft, New York.

Nicklas, R.B. (1975). In "Molecules and Cell Movement" (S. Inoué and R.E. Stephens, eds.), pp. 97-117. Raven Press, New York.

Nicklas, R.B., and Staehly, C.A. (1967). Chromosoma 21,1-16.

Niemark, H.E. (1977). Proc. natn. Acad. Sci. U.S.A. 74, 4041-4045.

Nishino, Y., and Watanabe, Y. (1977a). Biochim. biophys. Acta 490, 132-143.

Nishino, Y., and Watanabe, Y. (1977b). Biochim. biophys.

Acta 493, 104-114.

Olden, K., and Yamada, K.M. (1977). Analyt. Biochem. 78, 483-490.

Palevitz, B.A. (1976). In "Cell Motility. Book B: Actin, Myosin, and Associated Proteins" (R. Goldman, T. Pollard and J. Rosenbaum, eds.), pp. 601-611. Cold Spring Harbor Laboratory, Cold Spring Harbor, New York.

Pepper, D.A., and Brinkley, B.R. (1977). Chromosoma 60, 223-235.

Perry, M.M., John, H.A., and Thomas, N.S.T. (1971). Expl Cell Res. 65, 249-253.

Pollard, T.D. (1977). In "International Cell Biology 1976-1977" (B.R. Brinkley and K.R. Porter, eds.), pp. 378-387. Rockefeller University Press, New York.

Pollard, T.D., and Weihing, R.R. (1974). C.R.C. Crit. Rev. Biochem. 2, 1-65.

Pollard, T.D., Fujiwara, K., Handin, R., and Weiss, G. (1977). Ann. N.Y. Acad. Sci. 283, 218-236.

Pollard, T.D. Fujiwara, K., Niederman,R., and Maupin-Szamier, P. (1976). In "Cell Motility. Book B: Actin, Myosin and Associated Proteins" (R. Goldman, T. Pollard and J. Rosenbaum, eds.), pp. 689-724. Cold Spring Harbor Laboratory, Cold Spring Harbor, New York.

Porter, K.R. (1973). In "Locomotion of Tissue Cells, Ciba Foundation Symposium 14", pp. 150-169. Associated Scientific Publishers, Amsterdam, London, New York.

Puszkin, S., Kochwa, S., Puszkin, E.G., and Rosenfield, R.E. (1975). J. biol. Chem. 250, 2085-2094.

Ris, H. (1943). Biol. Bull. mar. biol. Lab., Woods Hole 85, 164-178.

Ris, H. (1949). Biol. Bull. mar. biol. Lab., Woods Hole 96, 90-106.

Robbins, E. (1961). J. biophys. biochem. Cytol. 11, 449-455.

Robbins, E., and Gonatas, N.K. (1964). J. Cell Biol. 21, 429-463.

Roos, U.-P. (1976). Chromosoma 54, 363-385.

Routledge, L.M., Amos, W.B., Yew, F.F., and Weis-Fogh, T. (1976). In "Cell Motility. Book A: Motility, Muscle and Non-muscle Cells" (R. Goldman, T. Pollard and J. Rosenbaum, eds.), pp. 93-113. Cold Spring Harbor Laboratory, Cold Spring Harbor, N.Y.

Ryser, U. (1970). Z. Zellforsch. 110, 108-130.

Sakai, H., Hiramoto, Y., and Kuriyama, R. (1975). Development, Growth, & Differentiation 17, 265-274.

Sakai, H., Mabuchi, I., Shimoda, S., Kuriyama, R., Ogawa, K., and Mohri, H. (1976). Development, Growth, & Differentiation 18, 211-219.

Sakai, H., Shimoda, S., and Hiramoto, Y. (1977). Expl Cell Res. 104, 457-461.

Sanger, J.W. (1975a). Proc. natn. Acad. Sci. U.S.A. 72, 1913-1916.

Sanger, J.W. (1975b). Proc. natn. Acad. Sci. U.S.A. 72, 2451-2455.

Sanger, J.W., and Sanger, J.M. (1976). In "Cell Motility. Book C: Microtubules and Related Proteins" (R. Goldman, T. Pollard and J. Rosenbaum, eds.), pp. 1295-1316. Cold Spring Harbor Laboratory, Cold Spring Harbor, N.Y.

Sato, H., Ellis, G.W., and Inoué, S. (1975). J. Cell Biol. 67, 501-517.

Sato, H., Ohnuki, Y., and Fujiwara, K. (1976). In "Cell Motility. Book A: Motility, Muscle and Non-muscle Cells" (R. Goldman, T. Pollard and J. Rosenbaum, eds.), pp. 419-433. Cold Spring Harbor Laboratory, Cold Spring Harbor, New York.

Schloss, J.A., Milsted, A., and Goldman, R.D. (1977). J.Cell Biol. 74, 794-815.

Schrader, F. (1953). Mitosis: the Movements of Chromosomes in Cell Division. Columbia Univ. Press (N.Y.).

Schroeder, T.E. (1973). Proc. natn. Acad. Sci. U.S.A 70, 1688-1692.

Schroeder, T.E. (1975). In "Molecules and Cell Movement" (S. Inoué and R.E. Stephens, eds.), pp. 305-334. Raven Press, New York.

Schroeder, T.E. (1976). In "Cell Motility, Book A: Motility, Muscle and Non-muscle Cells" (R. Goldman, T. Pollard and J. Rosenbaum, eds.), pp. 265-277. Cold Spring Harbor Laboratory, Cold Spring Harbor, New York.

Stephens, R.E. (1973). J. Cell Biol. 57, 133-147.

Stumph, W.E., Elgin, S.C.R., and Hood, L. (1974). J. Immun. 113, 1752-1756.

Swann, M.M. (1951a). J. exp. Biol. 28, 417-433.

Swann, M.M. (1951b). J. exp. Biol. 28, 434-444.

Taylor, D.L. (1976). In "Cell Motility. Book B: Actin, Myosin and Associated Proteins" (R. Goldman, T. Pollard and J. Rosenbaum, eds.), pp. 797-821. Cold Spring Harbor Laboratory, Cold Spring Harbor, New York.

Taylor, D.L., Condeelis, J., Moore, P., and Allen, R.D.
    (1973). J. Cell Biol. 59, 378-394.
Taylor, E.W. (1965). In "Proc. Fourth Intern. Congress on
    Rheology, Brown University" (E.H. Lee, ed.), pp. 175-191.
    Interscience, New York.
Tilney, L.G. (1975). In "Molecules and Cell Movement" (S.
    Inoué and R.E. Stephens, eds.),pp. 339-388. Raven Press,
    New York.
Tobin, S.L. and Laird, C.D. (1977). J. Cell Biol. 75, 150a.
Trenchev, P., and Holborow, E.J. (1976). Immunology 31,
    509-517.
Wang, E., Wolf, B.A., Lamb, R.A., Choppin, P.W., and Goldberg,
    A.R. (1976). In "Cell Motility, Book B: Actin, Myosin
    and Associated Proteins" (R. Goldman, T. Pollard and J.
    Rosenbaum, eds.), pp. 589-599. Cold Spring Harbor Lab-
    oratory, Cold Spring Harbor, New York.
Weber, K. (1976). In "Cell Motility. Book A: Motility,
    Muscle and Non-muscle Cells" (R. Goldman, T. Pollard and
    J. Rosenbaum, eds.), pp. 403-417. Cold Spring Harbor
    Laboratory, Cold Spring Harbor, New York.
Wolpert, L. (1965). Symp. Soc. gen. Microbiol. XV, 270-293.

EXPERIMENTAL STUDIES
OF MITOSIS IN THE FUNGI

I. Brent Heath

Biology Department
York University
Toronto, Ontario
Canada

I. INTRODUCTION

The purpose of this review is to examine aspects of mitosis in various fungi with the aim of explaining as much as is known about how fungi segregate their genomes into daughter nuclei. This mechanistic approach suggests that as broad a review as possible will be most useful hence the terms "experimental" and "fungi" will be used in their broadest sense, in fact only the following types of study will be excluded. The genetic analysis approach has great potential but because it has yet to provide substantial information about the fundamental mechanics of mitosis it has been excluded. Similarily much of the early light microscopy has been excluded, partly because it has been ably reviewed by Olive (1953) and Robinow and Bakerspigel (1965) and partly because much of it, but certainly not all, is of dubious value when viewed with the benefits of hindsight and newer techniques. The choice of organisms considered has been made on the basis of the range of taxa commonly considered in mycology courses rather than the recommendations of formal nomenclature. This choice means one is working with an undoubtedly polyphyletic group of organisms (Fuller, 1976). Finally, whilst concentrating on mitosis, data from meiotic nuclei will be included when it seems to make a contribution to the understanding of mitosis. Aspects of meiosis pertaining to recombination and haploidization will not be considered.

Two excellent recent reviews overlap with the present one. Kubai (1975) has considered much of the data reviewed here as an aid to understanding how the mitotic spindle evolved whilst Fuller (1976) has reviewed many of the ultrastructural observations using a semi phylogenetic framework. This work will' attempt to use a more component oriented framework in an effort to emphasize the similarities and differences found in various aspects of fungal mitosis as compared with the processes found in other organisms (reviewed by Forer in this volume). A similar, but less extensive, review along these lines has been previously published and contains the base of a number of points developed here (Heath, 1974a). As in that paper, much of the basic data considered in this paper will be presented in tabular form for more efficient access and referencing.

II.   NUCLEUS ASSOCIATED ORGANELLES (NAOs)

The general term nucleus associated organelle (NAO) was introduced by Girbardt and Hädrich (1975) to describe those structures which lie adjacent to, or within, the nuclear envelope at the spindle poles, or elsewhere during interphase, of many fungi. As discussed earlier by Girbardt (this volume), it has been adopted because it is functionally and morphologically neutral and is chronologically more accurate than Spindle Pole Body (Aist & Williams, 1972) since the structures referred to are usually associated with the nucleus throughout the nuclear cycle but only briefly lie at the spindle poles during mitosis or meiosis. The term is essentially synonymous with spindle pole body (SPB) (Aist & Williams, 1972), kinetochore equivalent (KCE) (Girbardt, 1971), microtubule organizing centre (MTOC) (Pickett-Heaps, 1969), archontosome (Beckett & Crawford, 1970), spindle or centrosomal plaque (Robinow & Marak, 1966) and nucleus associated body (NAB) (Roos, 1975a), all of which it replaces in this review. The term centriole will be retained for the well known triplet microtubule containing structure (Fulton, 1971) although perhaps formally it should be regarded as a sub-type of NAO.

A. Range of Morphology

As shown in Table 1, there are a number of generalizations which can fairly be made about the types of NAOs associated with fungal nuclei. Organisms which have a flagellate
stage in their life cycle generally have a pair of centrioles
associated with each nucleus. Centriole replication precedes
mitosis so that at each spindle pole there is usually a pair
of centrioles. The time of centriole replication is probably
variable. For example in Thraustotheca and Saprolegnia
(Heath, 1974b, Heath & Greenwood, 1970) it occurs immediately
before spindle formation; however, in those species where
centriole migration occurs before spindle formation it is
possible that replication and migration occur some long time
before mitosis. In Sapromyces (Heath, unpublished), for instance, it is very common to find interphase nuclei with replicated, migrated centrioles but no spindle suggesting that
this configuration is long lasting and thus indicating that
centriole replication occurs well before mitosis, possibly
close to the preceding telophase (cf. a similar phenomenon in
the Basidiomycetes [Girbardt & Hädrich, 1975, Heath & Heath,
1976]). Since, under favourable circumstances, the centriolar
region can be detected in living cells (McNitt, 1973), it
should be possible to directly determine the time of centriole
migration but this has not yet been achieved. Whatever the
time of replication, it is interesting to note that in at
least some Oomycetes, centriole behaviour mirrors the behaviour of the DNA by not replicating between meiosis I and II,
thus the haploid nuclei are each accompanied by a single centriole (Howard & Moore, 1970, Heath, unpublished). Such behaviour could be evidence for a common control mechanism but
such a mechanism cannot be universal since Perkins (1970) has
shown that, in Labyrinthula, centriole replication does occur
between the two meiotic divisions. The Oomycete system may
have evolved to ensure bi-parental inheritance of centrioles
but the value of such a system is totally obscure.
      Even in those organisms which do possess a flagellate
stage in their life cycles, centrioles are not always found
with the nuclei during all stages of the life cycle. For example they are absent from the plasmodial nuclei of the Myxomycetes even though they occur at the poles of myxamoebal
mitotic figures (Table 1). The situation is more unusual in
the Labyrinthulales where protocentrioles, which may contain

TABLE I.   Features of Fungal Mitoses

Organism	Class	Reference	Type of NAO
Labyrinthula	L	Perkins & Amon, 1969 Perkins, 1970 Porter, 1972	procentrioles and centrioles
Sorodiplophrys	L	Dykstra, 1976	procentrioles
Thraustochytrium	L	Kazama, 1974	centrioles
Cavostelium	P	Furtado & Olive, 1970	?
Polysphondelium	Ac	Roos, 1975	block
Dictyostelium	Ac	Moens, 1976	sphere
Physarum	Mp	Aldrich, 1967 Guttes, et al., 1968 Aldrich, 1969 Goodman & Ritter, 1969 Ryser, 1970 Sakai & Shigenaga, 1972 Tanaka, 1973	none
Echinostelium	Mp	Hinchee, 1976	none
Clastoderma	Mp	McManus & Roth, 1968	none
Didymium	Mp	Aldrich & Carroll, 1971	none
Arcyria	Mp	Mims, 1972	none
Ceratiomyxa	Mp	Scheetz, 1972	?
Physarum	Mm	Aldrich, 1969	centrioles
Sorosphaera	Pl	Braselton, et al.,1973,1975	centrioles
Polymyxa	Pl	Keskin, 1971	?
Plasmodiophora	Pl	Garber & Aist, 1977	centrioles

Time of NAO replication	Time of NAO or equivalent migration	Angle between centrioles	Nuclear envelope behaviour	Nucleolus behaviour	Metaphase plate	Chromosomal microtubules	Chromosome-to-pole shortening	Pole-to-pole elongation	Arrangement of non chromosomal microtubules	Loss of nucleoplasm
P	B	90 and 180	PF	D	?	+ 1?	?	+	d	?
P	B	90	I	D	?	?	?	+	d	?
P	B	90	PF	D	+	+ 1	+	+	d	+
?	?	?	D	D	+	+ ?	?	?	d	- ?
P	D	0	PF	p	-	+ 1	+	+	b	-
P	D	0	PF	p	-	+2 -3	+	+	b	-
0	?	0	PF	D	+	+1 -2	+	+	d	-
0	?	0	PF	D	+	+	+	+	d	?
0	?	0	I	?	+	+	?	?	?	?
0	?	0	I	D	+	+	?	?	?	?
0	?	0	Dt	D	+	+	+	-	d	?
?	?	?	I	?	+	?	?	?	?	?
?	B ?	90	D	D	+	+ 1?	+	+	d	-
?	B ?	180	PF	p	+	+	-	+	d	-
?	?	?	I ?	p	+	?	?	?	?	?
?	?	180	PF	p	+	+	?	?	d	?

TABLE I. (continued)

Saprolegnia	O	Heath & Greenwood,1968,1970 Howard & Moore, 1970	centrioles
Thraustotheca	O	Heath, 1974b	centrioles
Achlya	O	Ellzey, 1974	centrioles
Aphanomyces	O	Hoch & Mitchell, 1972 Heath, 1974c	centrioles
Albugo	O	Khan, 1976	centrioles
Phytophthora	O	Elsner, et al. 1970 Hemmes & Hohl, 1973	centrioles
Sapromyces	O	Heath, unpublished	centrioles
Apodachlya	O	Heath, unpublished	centrioles
Rhizidiomyces	H	Fuller & Reichle, 1965 Fuller, pers. comm.	centrioles
Harpochytrium	C	Whisler & Travland, 1973	centrioles
Phlyctochytrium	C	McNitt, 1973	centrioles
Entophlyctis	C	Powell, 1975	centrioles
Blastocladiella	C	Lessie & Lovett, 1968	centrioles
Catenaria	C	Ichida & Fuller, 1968	centrioles
Allomyces	C	Robinow & Bakerspigel, 1965 Turian & Oulevey, 1971 Olson, 1974a,b	centrioles
Mucor	Z	McCully & Robinow, 1973	none
Phycomyces	Z	Franke & Reau, 1973	none
Basidiobolus	Z	Tanaka, 1970 Sun & Bowen, 1972 Gull & Trinci, 1974	ring ?
Ancylistes	Z	Moorman, 1976	plaque
Conidiobolus	Z	Robinow, pers.comm.	none
Strongwellsea	Z	Humber, pers. comm.	ring
Lipomyces	A	Tanaka, 1977a & pers. comm.	plaque
Dipodascus	A	Tanaka, 1977b	plaque
Taphrina	A	Tanaka, 1977b	plaque
Podospora	A	Zickler, 1970, 1971	plaque
Ascobolus	A	Zickler, 1970, 1971 Wells, 1970	plaque

P	D	180	I	p	-	+ 1	+	+	d	-
P	D	180	I	p	-	+ 1	+	+	d	-
?	?	180	I	p	-	+	?	+	d	?
P	D ?	180	I	p	-	+ 1	+	+	d	-
?	?	180	I	p	-	+	+	?	d	?
P ?	D	180	I	p	-	+ 1?	+	+	d	-
T ?	B	180	I	p	-	+ 1	+	+	d	- ?
T ?	B	180	I	p ?	?	?	?	?	?	?
?	B	90	PF	D	+	+ 1	?	?	d	+
P	B	90	PF	D	+	+ 1?	+	+	d	-
P	B	90	PF	Di	+	+	+	+	d	+
P	B	90	PF	?	+	+	+	+	d	+
?	?	90	I	p	?	?	?	+	d	?
P ?	B	90	I	Di	+	+	+	+	d	+
?	?	90	I	Di	+	+	+	+	d	+
O	D	0	I	p	?	+ ?	+ ?	+	b	-
O	?	0	I	p	?	?	?	+	b	-
Pm	B	90	D	Df	+	+ 1?	+	+	d	-
P	B ?	0	I	p	?	+ ?	?	+	d	-
O	D	0	I	p	?	+ ?	?	+	b ?	?
?	D ?	90 ?	I	?	+	+	+	+	d	- ?
P	D	0	I	Di	-	+	+	+	b	+
?	?	0	I	Di	-	?	?	?	?	?
?	?	0	I	p ?	-	+	?	?	?	?
?	?	0	I	p	-	+	+	+	d	-
?	?	0	I	p	-	+ 5+	+	+	d	-

TABLE I. (continued)

Xylosphaera	A	Beckett & Crawford, 1970	plaque
Neurospora	A	Van Winkle, et al., 1971	sphere
Aspergillus	A	Robinow & Caten, 1969	plaque
Ceratocystis	A	**Stiers, 1976, Aist, 1969**	**plaque**
Pustularia	A	Schrantz, 1967	plaque
Erysiphe	A	McKeen, 1972	plaque
Saccharomyces	A	Moens & Rapport, 1971 Moor, 1966 Matile, et al., 1969 Robinow & Marak, 1966 Guth, et al., 1972 Peterson, et al., 1972 Byers & Goetsch, 1975a,b Zickler & Olson, 1975 Peterson & Ris, 1976	plaque
Schizosaccharomyces	A	McCully & Robinow, 1971	plaque
Fusarium	I	Aist & Williams, 1972	plaque
Cochliobolus	I	Huang, et al., 1975	plaque
Leucosporidium	Bh	McCully & Robinow, 1977b	sphere
Rhodosporidium	Bh	McCully & Robinow, 1972a	bar
Aessosporon	Bh	McCully & Robinow, 1972a	bar
Ustilago	Bh	Poon & Day, 1976a,b	dome-bar
Uromyces	Bh	Heath & Heath, 1976	plaque
Puccinia	Bh	Harder, 1976a,b	plaque
Gymnosporangium	Bh	Mims, 1977	plaque ?
Coprinus	B	Lu, 1967 Lerbs & Thielke, 1969 Lerbs, 1971 Raju & Lu, 1973 Thielke, 1974 Gull & Newsam, 1976	sphere
Poria	B	Setliff, et al., 1974	sphere
Armillaria	B	Motta, 1967, 1969	sphere
Boletus	B	McLaughlin, 1971	sphere
Polystictus	B	Girbardt, 1968 Girbardt & Hädrich, 1975	sphere
Schizophyllum	B	Sundberg, 1977	sphere
Phanerochaete	B	Setliff, 1977	?

?	?	O	I	p	-	+ 2+	?	?	?	?
P	?	O	I	p	-	+ 2+	?	+	b	-
?	D	O	I	?	-	?	?	+	b ?	- ?
?	D	O	I	?	-	?	+	+	b	+ ?
?	?	O	D ?	?	?	+	?	?	?	?
?	?	O	I	p	-	+ 2+	?	+	d	-
P	D	O	I	p	-	+ 1	+	+	b	-
P	D ?	O	I	p	?	?	?	+	b	-
?	?	O	I	D	-	+	+	+	b	- ?
P	?	O	I	Di	-	+	+	+	d ?	+
?	D	O	t	Di	?	+	?	+	b	+
P	D	O	t	Di	?	+ ?	?	+	b	+
P	D	O	t	Di	?	+ ?	?	+	b	+
P	D	O	Da	Di	-	+ 1	+ ?	+	d → b	+
T	D	O	I	Di	-	+ 3?	+ ?	+	b	+
?	?	O	I	Di	-	+	?	+	d → b	+
?	?	O	I	Di	-	?	?	?	?	+
P ?	D	O	D and I	D	-	+ 2+	+	+	d	?
?	?	O	PF	p	-	+	+	+	b	?
?	?	O	D ?	D	-	+	+	?	d	?
T ?	?	O	D	D	?	?	?	?	?	?
T	?	O	PF	?	?	+ ?	?	+	d ?	?
?	?	O	Da	?	?	?	?	?	b	?
?	?	O	PF	?	-	+	?	?	b	+

Explanation of Table 1. Throughout the table, + denotes presence of feature, - denotes absence of feature, ? denotes uncertainty and 0 denotes that that feature is not applicable to
that organism. Time of NAO replication; P = during prophase,
T = during telophase, Pm = during prometaphase. Time of NAO
or equivalent migration; B = before spindle formation, D =
during (or concomitant with) spindle formation. Angle between
centrioles; angle, in degrees, is only approximate. Nuclear
envelope behaviour; PF = polar fenestrae, I = remains intact,
D = disperses during prophase, Dt = disperses at telophase,
t = transiently opens then closes again, Da = disperses at
anaphase. Nucleolus behaviour; D = disperses during prophase,
p = persists throughout mitosis, Di = discarded, as a recognizable entity, into the cytoplasm, Df = becomes diffuse but is
detectable throughout mitosis. Metaphase plate; either present as a clearly defined plate (+) or the chromosomes lie variously spread along the length of the spindle (-). Chromosomal microtubules; either present (+) or unknown (?). When
present the number of microtubules per kinetochore is indicated. Chromosome-to-pole shortening, pole-to-pole elongation
and loss of nucleoplasm are all presence or absence features.
Arrangement of non chromosomal microtubules; these may form a
dispersed array (d) or may be aggregated to form a tightly
packed bundle (b). In those organisms for which a number of
investigations are cited, the best available data is given in
the table, by no means all features can be verified in each
paper. In some cases this author's judgement was used to
evaluate the presented data in the papers and the table en-
tries made according to that judgement.

only the cartwheels (Porter, 1970) or cartwheels plus singlet
microtubules (Dykstra, 1976), are formed de novo at each
mitosis and become undetectable during interphase. These examples show that a) centriole formation can be accomplished
easily from less organized (and thus less energy demanding)
precursors in supposedly primitive organisms and b) centrioles
are by no means necessary for mitosis (a point well made elsewhere [Pickett-Heaps, 1971]) in which case it is puzzling that
many other organisms, including mammals, spend energy to syn-
thesize centrioles for every mitotic division.
     The arrangement of centrioles is not uniform. Consist-
ently among the Oomycetes and Plasmodiophoromycetes (Table 1)
and occasionally in Labyrinthula (Perkins, 1970), the paired

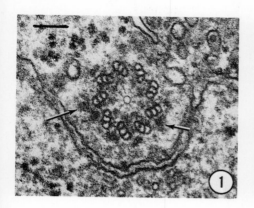 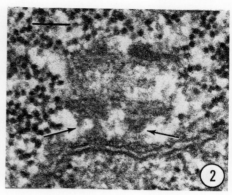

Fig. 1. <u>Aphanomyces</u> <u>astaci</u>. Transversely sectioned centriole showing the osmiophilic material (arrows) associated with the triplet microtubules and the zone of exclusion around the centriole. The centriole lies in a depression in the nuclear envelope. Bar = 0.1 μm. Unpublished.

Fig. 2. <u>Saprolegnia</u> <u>ferax</u>. A pair of longitudinally sectioned centrioles showing the prominent osmiophilic connections to the nuclear envelope (arrows). The hyphae were glycerinated during fixation, a procedure which enhances the visibility of the connections but otherwise gives a rather poor image. Bar = 0.1 μm. Unpublished.

centrioles lie end to end at 180° to each other whereas in most other groups they lie between 90° and parallel to one another (Table 1). To date no functional advantage has been envisaged for either arrangement hence its significance may be purely phylogenetic (Heath, 1975).

It must be emphasized that in all organisms possessing centrioles associated with the nuclei there is typically a variable amount of osmophilic material associated with the triplet structure itself, all of which is surrounded by a zone of exclusion of variable size (Fig. 1). As discussed later, the osmiophilic material may well be involved in microtubule polymerization. The zone of exclusion is totally uncharacterized, both as to composition and function. It may play a role in maintaining the close association between the centrioles and the nuclear envelope but there are more osmiophilic

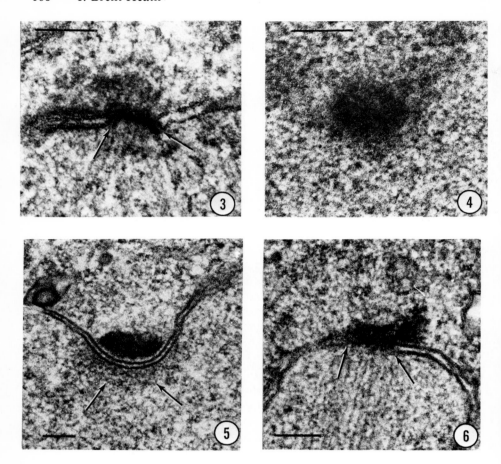

Fig. 3. <u>Saccharomyces</u> <u>cerevisiae</u>. Transverse section of
an NAO inserted in a pore of the nuclear envelope at the pole
of a mitotic spindle. Note the edges of the pore marked by
the arrows. Bar = 0.1 μm. Unpublished, courtesy of M.L.
Ashton, York University.

Fig. 4. <u>Saccharomyces</u> <u>cerevisiae</u>. Surface section of an
NAO comparable to that shown in Fig. 3 showing the circular
shape of the structure. Bar = 0.1 μm. Unpublished, courtesy
of M.L. Ashton, York University.

Fig. 5. <u>Schizosaccharomyces</u> <u>octosporus</u>. Transverse section of an interphase NAO showing an osmiophilic bar adjacent
to a depression in a clearly continuous nuclear envelope.

Note unidentified material adjacent to the NAO associated region of the nuclear envelope (arrows). Bar = 0.1 µm. Unpublished, courtesy of M.L. Ashton, York University.

Fig. 6. Schizosaccharomyces octosporus. One of a series of transverse sections of an NAO at the pole of a metaphase mitotic spindle. Note that the nuclear envelope is discontinuous adjacent to the NAO, the edges of the pore being indicated by the arrows. Bar = 0.1 µm. Unpublished, courtesy of M.L. Ashton, York University.

structures which appear more suited to this role (Fig. 2).

Among the organisms which lack centrioles there is considerable diversity in NAO morphology but to some extent these variations are conserved within major taxa (Table 1). For example among the Ascomycetes the NAO usually has a multilayered, osmiophilic, disc-like form which is closely appressed to the nuclear envelope (Figs. 3-6). Because of fixation difficulties it is often hard to resolve the interrelationship between the nuclear envelope and the ascomycete NAO but there does appear to be some variability. In Saccharomyces the NAO is permanently located in a close fitting hole in the nuclear envelope (e.g. Peterson, et al., 1972 and Fig. 3) but in other genera such as Erysiphe (McKeen, 1972) and Ascobolus (Wells, 1970, Zickler, 1970) it always lies on the continuous nuclear envelope. In Schizosaccharomyces it lies on the continuous nuclear envelope during interphase (Fig. 5) but fits into a pore in the envelope during mitosis (Fig. 6) (c.f. the rust Uromyces, Heath & Heath, 1976). The NAO may indeed become independent from the nuclear envelope, either in part, during mitosis in some genera (e.g. Wells, 1970, Zickler, 1970), or entirely when it appears to function in other roles after meiosis (Beckett & Crawford, 1970).

There is a variability in size and shape from the ± 0.1 µm diameter circle of Saccharomyces (Peterson, et al., 1972 and Fig. 4) to the 1-1.5 µm diameter, more oval, structure of Ascobolus (Zickler, 1970, Wells, 1970). In any one organism, apart from enlargement during replication, the size generally remains constant through interphase, mitosis and meiosis (e.g. Wells, 1970, Byers & Goetsch, 1975, Moens & Rapport, 1971) although there are exceptions to this. For example Zickler (1970) notes that in both Ascobolus and Podospora they get larger from prophase of meiosis I to post-meiotic mitosis

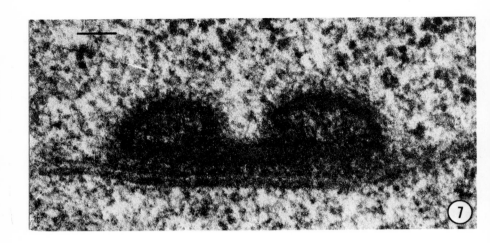

Fig. 7. <u>Boletus</u> <u>rubinellus</u>. Median longitudinal section
of an interphase NAO showing the two globules and interconnecting mid-piece. Bar = 0.1 μm. From McLaughlin (1971),
courtesy of Rockefeller University Press.

whilst Moens and Rapport (1971) showed them enlarging during
prospore wall formation at the end of meiosis II in <u>Sacchar-</u>
<u>omyces</u>. Whilst little survey type data is available it seems
that the species specific, as opposed to life cycle stage
specific, variation in size is probably a function of the
number of spindle microtubules which must associate with the
NAO (compare for example, <u>Saccharomyces</u> and <u>Ascobolus,</u> references above). This concept is supported by the report of
larger NAOs in diploid <u>vs</u> haploid yeast (Peterson & Ris,1976).
There are reported exceptions to the general disc-like
morphology of the ascomycete NAO. In a very brief report
Girbardt (1971) showed double, more globular structures in
<u>Aspergillus,</u> <u>Chaetomium</u> and <u>Neurospora</u>. A more detailed study
is needed to determine if these double structures were a repl-
icative phase but Künkel and Hädrich (1977) have shown more
detailed data on double <u>Neurospora</u> NAOs and Van Winkle et al.
(1971) also showed globular structures in <u>Neurospora</u> at various stages of mitosis. Thus it seems that one cannot yet
make broad generalizations about ascomycete NAO morphology.
NAO morphology and behaviour among the Basidiomycetes is
a little hard to describe in general terms because of reported

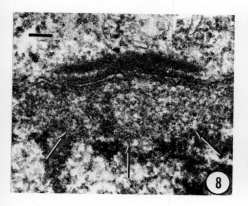 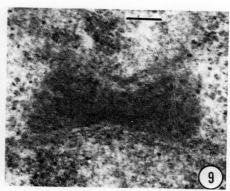

Figs. 8 & 9. Uromyces phaseoli var vignae. Median longitudinal (8) and surface (9) sections of interphase NAOs showing the two flattened, three layered, disc-like structures
with an interconnecting mid-piece. Note the hemisphere of
abundant granular material inside the nucleus adjacent to the
NAO (arrows in 8). Bars = 0.1 μm. From Heath & Heath (1976),
courtesy of Rockefeller University Press.

variations, not all of which are supported by good evidence.
Among the homobasidiomycetes and the yeast heterobasidiomycetes the interphase NAO is most commonly described as an
osmiophilic double globular or bar like structure with an interconnecting mid-piece between the globules (Fig. 7 and Table
1). This structure is normally closely associated with the
continuous nuclear envelope which frequently forms a deep depression in the region of association. In contrast to the
Ascomycetes, the basidiomycete NAO dissociates from the nuclear envelope so that the segregated globules or bars lie at
the poles of the mitotic or meiotic spindles, either free in
the nucleoplasm (e.g., McCully & Robinow, 1972a and b, Poon &
Day, 1976) or in large fenestrations of the nuclear envelope
at the nucleoplasm-cytoplasm interface (e.g. McLaughlin, 1971,
Girbardt & Hädrich, 1975, Setliff, et al., 1976, Lerbs &
Thiekle, 1969, Gull & Newsam, 1976) or in ball like extensions
of the nucleoplasm enclosed by an expanded nuclear envelope
(Thielke, 1974). In the homobasidiomycetes, but not the
heterobasidiomycetes, the globules enlarge considerably during
nuclear division (Girbardt & Hädrich, 1975, McLaughlin, 1971,

Setliff, et al., 1974). The functional significance of this enlargement is obscure but the reported correlation between nuclear size and NAO size in different species (Girbardt, 1971) may (assuming a correlation between spindle size and nuclear size) indicate that there is also a correlation between NAO size and spindle size as mentioned above for the Ascomycetes. The rust fungi present a variation on the above theme by also having a double interphase NAO, but in this case a double disc structure rather than double globules (Coffey, et al., 1972, Dunkle, et al., 1970, Harder, 1976a, Heath & Heath, 1976 and Figs. 8 & 9). The discs are more akin to ascomycete NAOs and their insertion into a close fitting hole in the nuclear envelope during mitosis is also more Ascomycete like, points which may have phylogenetic significance (Heath & Heath, 1976). In addition to variability in NAO morphology (i.e. discs vs. bars vs. globules) the basidiomycetes also appear to show variability in time of NAO replication. It is clear that the double structures are replicated NAOs. It is also clear, in Polystictus (Girbardt & Hädrich, 1975), Uromyces (Heath & Heath, 1976) and probably in the other species with double interphase NAOs mentioned above, that NAO replication occurs, via a sausage-shaped intermediate, shortly after telophase. Likewise there is good evidence for prophase NAO duplication in Ustilago (Poon & Day, 1976a). Ambiguity occurs in other species which have not been examined by serial sectioning. For example in Coprinus (Lu, 1976, Raju & Lu, 1973) and perhaps Poria (Setliff, et al., 1974), it is said that the pre-meiotic interphase NAO is a single globule with replication to a double globule occurring at prophase but such an assertion cannot be proven without serial sectioning. If this sequence is proven it may not mean that the interphase NAO in vegetive cells is also mono-globular because the above claims are based on pre-fusion nuclei. Just as centriole replication during meiosis can be controlled to give single centrioles in the Oomycetes (see above), so then the last mitosis which gives rise to the pre-fusion nuclei may also have an unusual NAO replication process in which replication is suppressed to give single pre-fusion NAOs. Clearly further work with serial section analysis is needed to determine the extent of variability occurring in the time of NAO replication among the Basidiomycetes.

Whilst the description of the ascomycete and basidiomycete NAOs has centred on the osmiophilic structures, it is

clear from most of the references in Table 1 that the osmiophilic components are surrounded by a variously large, often
ill defined, zone of exclusion. The composition and function
of this zone is totally obscure but its existence must be considered since it appears to be an integral part of the NAO
which clearly affects the interaction between anything from
the cytoplasm or nucleus and the osmiophilic part of the NAO.

The mitotic cycle associated changes noted above in the
basidiomycete NAOs are to some extent mirrored by the NAOs of
the cellular slime molds. In both Polysphondelium (Roos,
1975a and b) and Dictyostelium (Moens, 1976) the similar, and
so far unique, complex NAOs undergo both morphological and
size changes during the cell cycle. These changes are in part
associated with the replication of the organelle, which occurs
at telophase (Moens, 1976), but they probably also reflect
functional changes since the interphase NAO is the focus of a
radial array of cytoplasmic microtubules whereas during mitosis it is at the centre of a bidirectional array of mitotic and
astral (cytoplasmic) microtubules. As with the basidiomycete
NAOs, the cellular slime mold NAOs also function remotely from
the nuclear envelope, being located up to 0.7 μm away from the
nucleus in interphase (Moens, 1976) and lying in a large fenestration of the nuclear envelope during mitosis.

Just as the cellular slime molds have a unique NAO morphology, so apparently have the Trichomycetes. Reichle and
Lichtwardt (1972) have briefly described an interphase NAO in
Harpella. This NAO is a multi osmiophilic-layered structure
containing a semicircular array of 8-9 small tubules. They
present no three dimensional details nor any data on changes
during the nuclear cycle but its unique structure does seem to
be an example of the way in which NAO morphology can correlate
with other taxonomic features.

The zygomycetes present an interesting diversity of
structure at the poles of the nuclear spindles, the most remarkable situation being found in Phycomyces (Frank & Reau,
1973), Mucor (McCully & Robinow, 1973) and Conidiobolus (Robinow, personal communication) where there is no obvious structure outside the persistent nuclear envelope nor has any
structure comparable to an NAO been described during interphase. There is some slight amount of osmiophilic material
inside the nuclear envelope but this can hardly be called an
NAO and is probably most directly comparable to the intranuclear material described later for Physarum. Ancylistes

(Moorman, 1976) does show a single layered plaque-like structure whose morphology and behaviour appears to be most closely
similar to the NAO of Saccharomyces as described above. However the most intriguing type of NAO reported in a fungus
occurs in Basidiobolus where there is some uncertainty in the
reports of different authors. Sun and Bowen (1972) report a
centriole like structure which is cylindrical with a diameter
and length of 0.14 µm. They show no cross sections, but tubular elements can be discerned in the walls of the cylinder in
longitudinal sections and a hub like structure is also reported. The cylinder is smaller than a typical centriole (e.g.
0.19 µm diameter, 0.17 µm length in Thraustotheca [Heath,
1974b]) but it could be a procentriole with only singlet microtubules as found in some of the Labyrinthulales (Table 1).
If further analysis confirms this as a true centriole its presence in an organism which totally lacks a flagellate stage in
its life cycle is unique in cell biology. In the same species
Gull and Trinci (1974) show what may be a comparable structure, also an osmiophilic cylinder but with dimensions of 0.12
µm diameter, 0.11 µm length. They were unable to detect any
tubular components in the walls of this cylinder. If their
observations were to be confirmed, the amorphous cylinder
would be more akin to the somewhat larger (0.19 µm diameter,
0.07 µm length) ring shown in the red alga, Membranoptera, by
McDonald (1972). Clearly a more detailed account of the NAO
of Basidiobolus is needed; either reported structure, if confirmed, being somewhat unexpected and exciting in its possible
implications. Either way, these structures reinforce the impression obtained from the structure of the spindle (discussed
later) that Basidiobolus is rather distantly related to the
other zygomycetes, although another member of the Enteromoph-
thorales, Strongwellsea also has polar ring like structures
(Humber, personal communication). From a functional point of
view, it should be pointed out that in all other fungi the
spindle microtubules converge on the polar NAOs thus supporting the idea that the NAO has a role in microtubule formation
(discussed below) but in Basidiobolus where the spindle "poles" form a diffuse circular area with a diameter in the order
of 8 µm it is very hard to conceive of a similar function for
an organelle with a diameter of approximately only one six-
tieth of the spindle pole!

    In concluding a discussion of NAO morphology it must be
emphasized that whilst morphology may correlate with other

phylogenetic markers it becomes more obvious with the investigation of more species that increasing diversity within taxa
is likely to be found. Thus at present one should exercise
considerable caution in the use of NAO characteristics as phylogenetic markers.

B. Composition

     The composition of most NAOs, including centrioles, is
essentially unknown but considerable complexity is probably
the rule. Centrioles undoubtedly contain tubulin (the reader
is reminded that this is a generic term, there are undoubtedly
a number of closely related proteins to which this term applies [Snyder & McIntosh, 1976, Bibring, et al., 1976, Hepler
& Palevitz, 1974, Olmsted & Borisy, 1973]) and probably some
of the microtubule associated proteins found in other microtubules (Snyder & McIntosh, 1976). Since centrioles also contain cartwheels and other specific inter-microtubule links,
they must also contain other proteins which are presumably
centriole specific. In addition to the proteins there is some
evidence for the presence of both DNA and RNA in the analogous
basal bodies of ciliates (Randell & Disbrey, 1965, Smith-
Sonneborn & Plaut, 1976, Wolfe, 1972, Fulton, 1971) but the
data supporting the presence of DNA has been questioned and it
now seems probable that RNA is the only nucleic acid present
(Younger, et al., 1972, Hartman, et al., 1974, Dipell, 1976,
Heideman, et al., 1977). It is probable that RNA is also present in animal centrioles (Zackroff, et al., 1975) but it is
quite possible that the RNA may reside in the osmiophilic material surrounding the centriole rather than being an integral
part of the centriole itself.
     Among the non-centriolar type of NAOs found in the fungi
there is very little unambiguous compositional data. The most
complete information is for Ascobolus where Zickler (1973)
used electron microscope observations of enzyme treated cells,
and Feulgen staining, to show that the NAOs contained protein, probably basic protein, and DNA. RNAase gave insignifi-
cant structural changes indicating the presence of little, if
any, RNA. In Ustilago, Poon and Day (1974) used acridine
orange staining to show that the NAO fluoresced DNAase sensitive green during division and RNAase sensitive orange-red
during interphase. Such data is not unambiguous since the

authors showed that the quantity and colour of fluorescence is
affected by a number of experimental and cellular (cell cycle)
factors. Hartman, et al. (1974) have also pointed out a number of difficulties in interpreting acridine orange fluorescent
observations. Beyond these papers, published data is absent.
However, Girbardt (1977) has noted that the globular elements
of the Polystictus NAO give a DNA type reaction after selective extraction by EDTA whereas Borisy (1977) has reported that
the microtubule polymerization promotion activity of yeast
NAOs (Borisy, et al., 1975) is insensitive to both DNAase and
RNAase. These reports cannot be evaluated but they clearly
indicate that much more work is needed to determine a) if
there are indeed any nucleic acids in fungal NAOs and b) what,
if any, role these nucleic acids have.

C. Functions

It seems highly probable that the primary function of
most fungal NAOs is to play some controlling role in microtubule polymerization. The simplest evidence for this role is
the morphological observations of both spindle and cytoplasmic
microtubules converging onto, and frequently joining, the NAO
in most of the cells referred to in Table 1. Stronger evidence comes from cells such as Rhodosporidium (McCulley & Robinow, 1972a), and the homobasidiomycetes referred to in Table
1, where the entry of the NAO into the nucleus is a prerequisite for spindle formation. Further evidence is provided by
the chytrids (Table 1) in which spindle microtubules grow from
the centriolar region and rupture the nuclear envelope as they
form the spindle. Likewise the double divergent array of
spindle microtubules radiating from the migrating prophase
NAOs of Saccharomyces (Moens & Rapport, 1971, Peterson & Ris,
1976, Byers & Goetsch, 1975) strongly suggests a NAO role in
organizing spindle microtubules. Recently more direct experimental evidence for this role has been obtained by Borisy, et
al. (1975) who obtained in vitro microtubule polymerization
onto isolated yeast NAOs. However Setliff (1977) has recently
reported that the interphase NAO of Phanerochaete does not
become the structure upon which the spindle microtubules converge at the spindle poles. The interphase NAO is apparently
displaced to one side by a larger structure which is formed
de novo at mitosis. This behaviour is similar to the

situation found in the diatom <u>Surirella</u> (Tippit & Pickett-Heaps, 1977) and emphasizes that there may well be diversity in behaviour and function of NAOs.

The role of centrioles in microtubule polymerization (excluding flagella axonemes) is more uncertain and may be a matter of semantics. Direct connections between microtubules (spindle and cytoplasmic) and the triplets and assorted inter-triplet linkages has been claimed (McNitt, 1976 and references therein) but, in this reviewer's opinion, not convincingly proven (it is hard to prove such connections given the problems of section thickness, small structures, and the possibility of superimposition). However it is clear in most of the papers listed as having centrioles in Table 1 that microtubules insert into the osmiophilic material which frequently surrounds centrioles. Experimental evidence for a role of this material in microtubule polymerization is seen in the recent work of Gould and Borisy (1977) where the pericentriolar osmiophilic material surrounding mammalian centrioles will promote microtubule assembly when separated from the centrioles. A superficially similar type of material from the flagellar roots of the alga <u>Polytomella</u> also acts as in initiator of microtubule assembly <u>in vitro</u> (Stearns & Brown, 1976). Thus it seems probable that the pericentriolar osmiophilic material of the fungi will be found to have microtubule assembly promoting properties. On ultrastructural grounds (which are not very solid!) the osmiophilic component of all fungal NAOs is comparable to this pericentriolar material. Thus it may well be that this material is chemically and functionally similar in all fungi and that its morphological organization varies in a way which may have a dependence on phylogeny as well as function (i.e. number and disposition of microtubules formed). In those organisms which, for obscure reasons, choose to retain centrioles with the vegetive nuclei, the osmiophilic material happens to have an affinity for, and thus is arranged around, the centrioles. As discussed by Pickett-Heaps (1971) the centriole thus has no role in spindle (nor cytoplasmic) microtubule formation and could be said to be merely going along for the ride during mitosis. The real microtubule organizing centre (MTOC, Pickett-Heaps, 1969) of these cells would thus be the above mentioned osmiophilic material.

The osmiophilic component of fungal NAOs and centrioles may well be an MTOC as discussed above but it is clear that

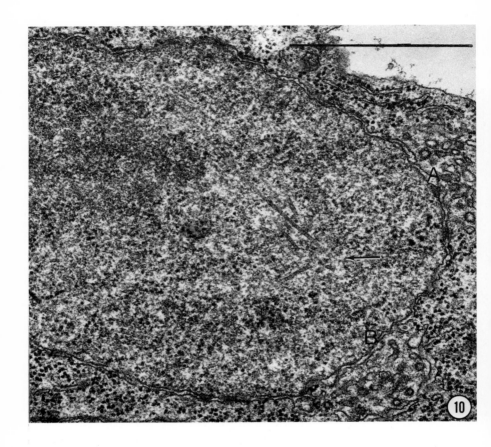

Fig. 10. <u>Sapromyces</u> <u>elongatus</u>. Presumed early prophase
nucleus showing microtubules converging on an undifferentiated
region of the nucleoplasm (arrow) remote from the nuclear envelope.   Serial sections confirmed a) that all nuclear microtubules converged on this region, b) the nuclear envelope came
nowhere near the region and c) that centriole pairs were located in the vicinity of the "A", 30 sections above this one
and at "B", 23 sections below this.   Bar = 1 μm.   Unpublished.

the mechanisms by which it works are highly complex.   For example, in <u>Uromyces</u> a single NAO can nucleate spindle microtubules on one side with very few cytoplasmic microtubules
during early phases of mitosis and subsequently nucleate far
more cytoplasmic microtubules during telophase (Heath & Heath,

1976). Similarly, during budding and conjugation in Sacchar-omyces, the arrangement of the NAO associated cytoplasmic microtubules undergoes a number of changes with no detected morphological changes in the NAO itself (Byers & Goetsch, 1975). An additional complication to understanding the microtubule organizing role of fungal NAOs, or any other MTOCs for that matter, is the mechanism by which the microtubules polymerize. The only clear case in which the site of microtubule polymerization is known is in flagella where distal addition of subunits to the forming microtubules is probable (Witman, 1973). If this observation is applicable to cytoplasmic and spindle microtubules then, apart from determining initial direction and start of synthesis, the way in which a basal NAO controls microtubule polymerization is obscure.

The final point concerning NAOs and MTOCs is that whilst the NAOs, etc. may well be primarily concerned with microtubule polymerization there are a number of cases in which other distinct entities appear to have the MTOC role for spindle formation. The best example is seen in Physarum in which an amorphous osmiophilic intranuclear structure, not associated with the nuclear envelope, forms the focal point of microtubule polymerization (Sakai & Shigenaga, 1972, Tanaka, 1973). A comparable situation appears to exist in the Oomycete, Sapromyces where microtubules can be found radiating from intranuclear foci which are also remote from the nuclear envelope and centrioles (Fig. 10, Heath, unpublished). Because microtubules do not traverse the nuclear envelope it is to be expected that all organisms in which the nuclear envelope remains intact during mitosis will eventually be shown to possess intranuclear MTOCs of some type. The morphology of such structures need not be complex, for example, the very thin layer of osmiophilic material adjacent to the nuclear envelope at the spindle poles of most intranuclear spindles probably represents the MTOC for the spindle microtubules of these organisms. (See also the polar structures of Phycomyces [Franke & Reau, 1973]). Since the pericentriolar material, when present, apparently represents an extranuclear MTOC, these organisms presumably have at least two MTOCs located at the polar regions of the spindle. As speculated by Fuller (1976) it may very well be that in some Ascomycetes and Basidiomycetes (those with the NAO in a pore or entering the nucleus), these two MTOCs have become fused into one organelle, the NAO, thus explaining the location of these organelles within

the nuclear envelope, or within variously large pores of the
nuclear envelope, during mitosis as discussed above. Presumably a similar unification of MTOCs has occurred in those organisms in which polar fenestrae are formed adjacent to the
centrioles during spindle formation (Table 1 and Section V,A);
in these cases the pericentriolar material would represent the
fused MTOCs. It should be emphasized that the use of the term
fused in this context is not intended to differentiate between
exclusive takeover of both MTOC functions by one MTOC of a
hypothetical preexisting pair versus actual fusion of the two
into one structure during evolution. Finally the reader is
reminded that the kinetochores may also have a MTOC role as
discussed later (Section V,A).

III. NUCLEAR ENVELOPE

     The obvious major feature concerning the nuclear envelope
during mitosis is the extent to which it breaks down. The
range of behaviour is shown in Table 1 from which it can be
seen that there is, in general, some homogeneity of behaviour
within the major taxa. Again the Myxomycetes illustrate a
major point. The behaviour of the nuclear envelope is not
constant within a given species, it may vary from completely
dispersing (i.e. an open spindle) to only having polar fenestrae (i.e. a closed spindle) within a single species (Aldrich, 1969) thus indicating that control of nuclear envelope
behaviour is more complex than being a simple characteristic
of the genetic properties of one organism. What else interacts with the genome to cause the different nuclear envelope
behaviour, and why the behaviour differs, is totally obscure.
It has been suggested that in coenocytes the retention of an
intact nuclear envelope is helpful in ensuring that there is
no fusion of spindles with the consequent possibility of genetically unbalanced progeny (Pickett-Heaps, 1974). However
this suggestion is rendered somewhat less convincing by the
observation of open spindles in many coenocytic amoebae (Ross,
1968, Goode, 1975). A more convincing explanation for the
trend to open spindles is to allow easier and more rapid access to tubulin and or microtubules from their probable site
of synthesis in the cytoplasm (e.g. Jorgensen & Heywood, 1974)
to the site of the spindle (Pickett-Heaps, 1972, Goode, 1975).
Such a hypothesis is hard to test and at present there is no

solid information for or against it. However Kerr (1967) did
show that the open spindles of Didymium work faster (i.e.
mitosis is quicker) than the closed ones of the plasmodia. If
all tubulin is synthesized in the cytoplasm, clearly it is
able to traverse the nuclear envelope in many organisms but
the degree to which such a process is slowed relative to ingress in the absence of an envelope, and the consequences of
any such hypothetical delay, are obscure. Since the trend
among organisms is towards open spindles as found in higher
plants and animals, one suspects that there is some functional
pressure driving evolution in that direction and ease of entry
of tubulin could be that pressure. However that does not answer the consequential question of what pressure has caused
retention of the intact envelope in many organisms. Undoubtedly the intact envelope would make it easier for a cell to
generate and maintain different ionic conditions inside the
nucleus versus the cytoplasm. In organisms which do not necessarily stop growing during division and in which there is an
obvious gradient of cytoplasmic properties from tip to sub-
apical regions (such as many fungi e.g. Zalokar, 1959), it may
well be advantageous to retain an intact nuclear envelope to
isolate the nuclear cycle of events (including mitosis) from
the cytoplasm. There is at least some evidence, in various
cells, to show that the nuclear envelope can apparently effect
control over the passage of various molecules and ions (e.g.
Feldherr, 1972, Goldstein, 1974). However it becomes dubious
if such can occur in those species which have polar fenestrae
(Table 1). With the advent of the microprobe X-ray analysis
system for electron microscopy it should be possible to obtain
more definite data on levels of various ions in nuclei at
different times of the nuclear cycle as compared with cytoplasmic levels, but to this author's knowledge no such data is
available for any fungus. Until more data are available, further speculation on the significance of the nuclear envelope
configuration is merely adding surfeit to excess. A similar
comment can probably be fairly made about the perinuclear endoplasmic reticulum which is present around the mitotic nuclei
of a number of classes of fungi (e.g. McNitt, 1973, Powell,
1975, Setliff, et al., 1974, Garber & Aist, 1977). Again the
functional significance of this material is unknown.

As seen in Table 1, and mentioned above, the behaviour of
the nuclear envelope in most major taxa is reasonably consistent. However one must always be aware of experimental error

in interpreting this behaviour. In general terms, those organisms listed as having an intact membrane, with or without
polar fenestrae, probably do show that feature in vivo although it must be admitted that the critical proof of serial
sectioning entirely through nuclei at various mitotic stages
is lacking in most cases. Great ambiguity exists in the case
of organisms listed as having nuclear envelopes which disperse
during mitosis. It is easy to envisage artifactual nuclear
envelope breakdown during fixation. The Basidiomycetes seem
to be the group in which most variability is reported; for example, Gull and Newsam (1976) illustrate metaphase meiotic
nuclei with the nuclear envelope ranging from intact with
small polar fenestrae to completely dispersed, all in one
species fixed in the same way. Likewise in another species of
Coprinus, Lerbs (1971) shows completely dispersed nuclear envelopes at metaphase whilst Thielke (1974) illustrates metaphases with the nuclear envelope intact and enclosing the
NAOs. Apart from the obvious potential for artifactual disruption of a continuous nuclear envelope during fixation and
for in vivo cell-cell variability, the possibility also exists
for mitotic stage dependent changes in nuclear envelope integ-
rity. Certainly the enclosing membrane around the NAOs shown
by Thielke (1974) could not coexist with the extensive cytoplasmic microtubules radiating from the NAOs shown by Lerbs
(1971) or Lerbs and Thielke (1969). A case for transient
opening and reclosing of the nuclear envelope during mitotic
prophase has been made for some heterobasidiomycetous yeasts
where the nuclear envelope opens up to admit the NAOs prior to
spindle formation then recloses for the rest of mitosis
(McCulley & Robinow, 1972a). Whatever the relative extent of
artifact versus real-but-transient dispersal of the nuclear
envelope proves to be, it is clear that the nuclear envelope
of the Basidiomycetes is more labile in its behaviour than
that of many of the other fungi. The reason(s) for this lability are totally obscure but it is clear that further study
with various different fixation regimes would be very helpful
in clarifying real versus artifactual lability.

Apart from the rather gross extremes of nuclear envelope
behaviour discussed above (i.e. completely open or dispersed
versus completely closed or intact) there are a number of more
subtle variations which deserve some mention.

The necessity for polar fenestrae in those species which
develop at least some of their microtubules from extranuclear

NAOs (e.g. the chytrids, Table 1) is self evident and in the myxomycete plasmodia where such a function is not needed early in mitosis it is possible that the fenestrae which form later in mitosis are important to allow some spindle microtubules to interact with the cytoplasm (e.g. Aldrich, 1969). Whilst production of polar fenestrae could well be due to mechanical stress induced by the growing microtubules (McNitt, 1973, Whisler & Travland, 1973) it is evident that the nuclear envelope is differentiated to some extent, otherwise, if it behaved as a simple skin over the nucleoplasm, the stress would result in random tears in non-polar regions. Similar differentiation is seen morphologically in the "pocket" (Heath & Greenwood, 1968, 1970) region of the nuclear envelope (i.e. the region adjacent to the centrioles or NAOs) in most species with intact nuclear envelopes (Table 1). Likewise, as pointed out by Fuller (1976), the apparent initiation of reformation of the nuclear envelope in the NAO region of the Armillaria nucleus (Motta, 1969) could instead be evidence for greater stability of what was in fact the in vivo intact nuclear envelope in this region. Further evidence for differentiation of various regions of the nuclear envelope is seen in the selective opening of regions of the envelope to permit NAO ingress in some basidiomycetous yeasts (McCully & Robinow, 1972a,b). Additional complexity is ascribed to the nuclear envelope by McCully and Robinow (1971) when they suggest that, by selective localized growth, the nuclear envelope helps separate the NAOs during spindle development in Schizosaccharomyces. Also, in Saprolegnia, the membrane associated kinetochores are sorted into two groups when the "pockets" replicate (Heath, unpublished and Fig. 11). Similarly, at the opposite end of mitosis, the way in which the nuclear envelope encloses the chromosome masses with exclusion of a mid-body in, for example, Catenaria (Ichida & Fuller, 1968) versus simple median constriction in the Oomycetes (Table 1) suggests complex membrane properties. However in all of these situations it is important to recall that one may not be dealing with membrane properties per se but rather with membrane associated material, such as the cell-membrane-associated actin network found in Dictyostelium cells (Clarke, et al., 1975). Furthermore, at least for most of the above examples of membrane behaviour, the nuclear envelope is in direct contact with the nucleoplasm which may itself have structure and properties (e.g. Berezney & Coffey, 1977) which could influence the nuclear envelope.

Thus the conclusion one should remember is that whilst the
nuclear envelope may exhibit various types of behaviour this
behaviour may well not be due to the membranes themselves but
to other associated and so far little studied material.

From a functional viewpoint one of the most intriguing
types of behaviour of the nuclear envelope may be that shown
in Sorosphaera (Braselton, et al., 1975). During division the
chromosome masses are covered on their poleward sides by fenestrated cisternae, portions of which are continuous with the
inner nuclear envelope membrane. Such continuity suggests an
origin of these cisternae from the nuclear envelope. Their
function is of course obscure but they would be excellent can-
didates for a sarcoplasmic reticulum-like role in ion control
during mitosis, a concept discussed in more detail later (Section IX). Whether the elaborations of the nuclear envelope
shown in Entophlyctis (Powell, 1975) have a comparable role is
entirely unknown but they do not associate with the chromosomes. They do form the new nuclear envelope at telophase but
this may not be their sole function.

Clearly the nuclear envelope exhibits diverse behaviour
during mitosis in the various fungi. The functional significance of this behaviour is obscure in most instances, thus
tempting one to believe that the behaviour may have phylogenetic value. However, the dimorphic behaviour of the Myxomycetes is a clear indication that caution is needed in this
respect until much more is known about the role of the nuclear
envelope in the biology of the cell.

IV. NUCLEOLI

In general terms, as seen in Table 1, the nucleolus of
various fungi can have essentially three fates during mitosis.
It may disperse at various times, usually during prophase or
metaphase, and thus be no longer recognizable in either the
nucleus or the cell. It may persist with no obvious reported
changes and merely become constricted into two at telophase
(fission) so that each daughter nucleus receives half of the
parental nucleolus. Finally it may be discarded into the
cytoplasm as a morphologically intact entity. When discarded
it may be associated with a substantial quantity of nucleoplasm and nuclear envelope. The time at which it is discarded
varies; in the uniflagellate fungi (Table 1) it is typically

discarded at telophase when the nuclear envelope constricts about the chromosomal masses and thus excludes a large nucleolus containing mid-piece (e.g. Ichida & Fuller, 1968). An essentially similar process seems to occur in Uromyces (Heath & Heath, 1976), Cochliobolus (Huang, et al., 1975), and possibly some other genera (Fuller, 1976) but Harder (1976a and b) proposes, but could not prove, that premitotic exclusion occurs in Puccinia species. However in the basidiomycetous yeasts and Ustilago (McCully & Robinow, 1972a,b, Poon & Day, 1976a,b) nucleolus exclusion does occur during prophase or early metaphase. (See also Robinow & Bakerspigel, 1965, for a list of organisms in which light microscope data indicates discarding of the nucleoli.) In all cases where nucleolar extrusion occurs the excluded material is presumably returned to the cellular biochemical pools, apparently fairly rapidly (Ichida & Fuller, 1968).

Although data is lacking, it is not unlikely that dispersion of the nucleolus merely indicates transient loss of activity with consequent loss of some components through routine export from the nucleus (e.g. ribosome components) and reduced visibility and or dispersion of the other components. Nucleolar fission might be expected to result if activity were not shut off during mitosis. However the discarding of the apparently entire structure to the cytoplasm is hard to comprehend, as yet no reasonable hypothesis has been postulated to explain this behaviour. Further discussion of the nucleolus is not relevant to this paper since, contrary to earlier statements (Robinow & Bakerspigel, 1965, p.132), there is no convincing evidence for an active role of the nucleolus in any mitotic system, although if present, it undoubtedly has some impact on the mechanisms of the process.

V.   SPINDLES

The essential function of mitosis is to ensure reliable equipartitioning of complete genomes to daughter nuclei. Contrary to earlier suggestions (Moore, 1964) based on what are now known to be inadequate techniques, there is no reason to believe that mitosis in any fungus occurs in the absence of some form of microtubule containing spindle. The details of the variations found in these spindles are the subject of this section. For convenience it is useful to here introduce the terminology used for the most obvious spindle components, the microtubules. Heath (1974b) described 5 types of microtubule, the terminology for which are as follows, a) <u>chromosomal mic-rotubules</u> which run from the chromosomes, or kinetochores, usually to the spindle poles, b) <u>continuous microtubules</u> which run from pole to pole, c) <u>interdigitating microtubules</u> which run from one pole to some point beyond the equator of the spindle and which therefore interdigitate with similar micro-tubules from the opposite pole, d) <u>polar microtubules</u> which run from one pole for a short distance and terminate in no identifiable structure in the nucleoplasm, and e) <u>free micro-tubules</u> (McIntosh, et al., 1975) which lie at various posi-tions in the spindle with no evident connection to either the poles or kinetochores.

A. Formation

    Detailed studies on spindle formation in the fungi are lamentably few; most reports begin at metaphase. The reasons for this are probably threefold: a) The pre-metaphase mitotic stages may occur very rapidly and thus be rarely encountered in fixed material. b) They are hard to unambiguously recog-nize in material which has not been serially sectioned. c) In many organisms the changes which occur during spindle forma-tion are at, or below, the limits of resolution of the light microscope and thus are hard to observe. Fortunately in at least some fungi, notably the Oomycetes, Acrasiomycetes and perhaps <u>Saccharomyces</u> (refs. in Table 1), premitotic states occur frequently in fixed material and can thus be easily des-cribed in some detail.
    As seen in Table 1, there are basically two models by which spindle formation in the fungi can occur. The NAOs may migrate and establish the future spindle poles at some consid-erable distance apart before there is any formation of the spindle, or the spindle may form between the migrating NAOs, elongating as they separate. In the former case, as discussed in Section II C, and perhaps best illustrated by McNitt (1973) and Whisler and Travland (1973), at least some of the spindle microtubules form by growing away from the centriolar region and entering the nucleus via the polar fenestrae. Which mic-rotubules form this way is unknown; it is highly probable that the continuous and interdigitating microtubules form from the

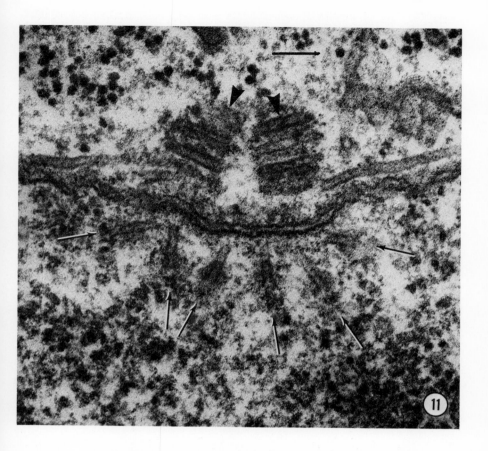

Fig. 11. <u>Saprolegnia</u> <u>ferax</u>. Early prophase. 6 of the
38 kinetochores (arrows) are shown forming a single group adjacent to the, as yet unreplicated, pair of centrioles (large
arrows). Bar = 0.1 µm. Unpublished.

centriolar region but the origin of the chromosomal microtubules is more problematic with growth from the poles to the
chromosomes or vice versa being equally possible. Very care-
ful serial section analysis of early prophase will be essential to resolve this question; at least in fungi with relatively small nuclei and spindles such a study is highly feasible.
   In organisms in which spindle formation is coincident
with NAO migration (Table 1) more detailed data is available
but still the direction of growth of, and site of addition of,

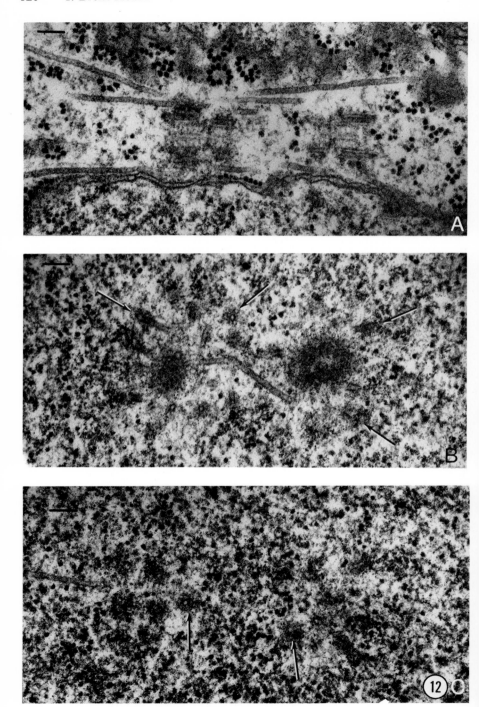

Fig. 12. <u>Saprolegnia ferax</u>. Early prophase. Centriole replication has occurred to produce two pairs of centrioles (A). Inside the nucleus, adjacent to each pair of centrioles, there is a hemispherical array of kinetochores, sections of which are seen in B and C. These arrays form two distinct but slightly overlapping groups, that on the left contained 22 kinetochores and that on the right 23. Representative kinetochores are arrowed. Note absence of any evidence for pairing between kinetochores of adjacent groups. Sections A, B and C were numbers 10, 5 and 2 in the series. Bars = 0.1 µm. Unpublished.

subunits to microtubules is obscure. The most detailed studies of the very early events in spindle formation have been performed on <u>Polysphondelium</u> (Roos, 1975), <u>Dictyostelium</u> (Moens, 1976), <u>Saccharomyces</u> (Peterson & Ris, 1976), <u>Thraustotheca</u> (Heath, 1974b) and <u>Saprolegnia</u> (Heath & Greenwood, 1970, Heath, unpublished and Figs. 11-13). In each of these organisms the non-chromosomal microtubules develop as short microtubules radiating in two groups, either from the separating NAOs or from the pocket regions of the nuclear envelope adjacent to the centrioles (depending on the species). In the yeast, the two groups of microtubules develop parallel to each other and reach their metaphase lengths before they reorientate to form continuous microtubules which lie between the now opposing NAOs, but in the other species many of them are oriented between the polar regions before they attain their metaphase lengths. In all of the above species the continuation of mitosis involves increasing interpolar distances with concomitant non-chromosomal microtubule elongation and anaphase movement of the kinetochores and chromosomes to the poles as discussed later. Because, at least in yeast, and probably in the other species also, the non-chromosomal microtubules do form to their metaphase lengths from two separate NAO regions, it is highly likely that they do indeed have opposite polarity when arranged in the mature spindle, an observation the significance of which will be more obvious in Section V E. There is more apparent variability in the formation of the chromosomal microtubules. In <u>Dictyostelium</u> the kinetochores replicate at telophase and are present as opposed pairs on each replicated chromosome at prophase. Microtubules do not form on these kinetochores at as early a stage as the non-chromosomal microtubules develop from the NAOs. It is not

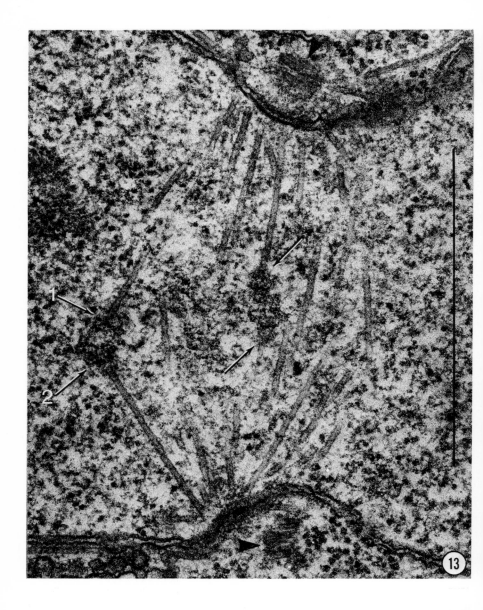

Fig. 13. <u>Sparolegnia ferax</u>. Early metaphase. Centrioles (large arrows) lie outside the nuclear envelope at the
spindle poles. Kinetochores (arrows), some of which appear to
be paired (e.g. arrows 1 and 2), lie in the spindle. Bar =
1 μm. Unpublished.

known whether they grow from pole to kinetochore or vice versa when they do form by metaphase. In contrast to Dictyostelium, the kinetochores of Saprolegnia replicate at prophase (Fig. 11), at which time they show no evidence of being paired (in the present work paired is used in the sense of two oppositely facing kinetochores as are found on mitotic chromosomes of higher organisms). They are all attached to one pocket region of the nuclear envelope (or more accurately to osmiophilic material immediately adjacent to the nuclear envelope; no direct microtubule-membrane connection occurs). Whilst apparently retaining this attachment they separate into two groups (Fig. 12) and later, as the spindle elongates, many, but perhaps not all, appear to pair (Fig. 13), in which case each member of a pair is attached to the opposite pole to its mate by a microtubule which first elongates to its metaphase length then shortens again during anaphase. (The term "appear to pair" is used above with caution because, whilst pairs undoubtedly exist (Fig. 13), it is possible that this configuration is a chance occurrence due to space constraints within the spindle. It may thus not indicate paired sister chromatids as are found in mitotic spindles of "higher" organisms.) This behaviour illustrates an important point which is also seen in Saccharomyces and possibly Polysphondelium, namely that separation of the genome into two distinct groups occurs at the very onset of mitosis prior to construction of the full spindle (assuming that the kinetochores are attached to the chromosomes, a highly probable but unproven situation, see below), a point to be returned to later (Section V E). Again, one cannot determine in which direction the kinetochore microtubules are polymerized because from their very inception they are "connected" to both the nuclear envelope and the kinetochore. The other studies in which spindle formation is coincident with NAO separation suffer from various technical problems which preclude them from contributing to the points discussed above.

One obvious feature of spindle formation which requires explanation is the mechanism by which the NAOs are moved apart during spindle development. Once a spindle exists between the separating NAOs, the elongation of this spindle very possibly moves the pole associated structures (centrioles, etc.) apart but obviously such a system cannot work for those organisms in which NAO migration precedes spindle formation and even in the other species there is a time during which separation occurs

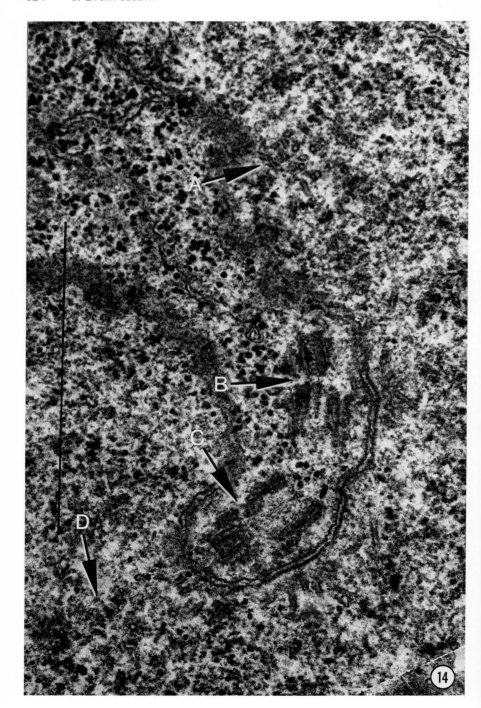

Fig. 14. <u>Saprolegnia ferax</u>. Nucleus and centrioles in a hypha treated with approximately 8.5×10^{-6} M oncodazole for 24 h. The nucleus was grossly enlarged and sections serial to this one revealed a total of 4 pairs of adjacent centrioles (2 of which are shown by large arrows B and C and the positions of the other 2 indicated by large arrows A and D). Each centriole pair was adjacent to an array of kinetochore microtubules each of which contained an approximately normal complement of kinetochores. Cytoplasmic microtubules were almost entirely absent. Bar = 1 μm. Unpublished.

before the microtubules are organized between the spindle poles. There are basically two mechanisms which have been postulated to explain this movement, a) membrane growth (McCully & Robinow, 1971) and b) cytoplasmic microtubule growth (Heath, 1974b). There is no unequivocal evidence to distinguish between these possibilities. In <u>Thraustotheca</u> (Heath, 1974b) cytoplasmic microtubules do form between the separating centrioles and their elongation is a good candidate for moving the centrioles as is also true in <u>Dictyostelium</u> (Moens, 1976) and <u>Polysphondelium</u> (Roos, 1975) where the nuclear envelope is absent between the initially separating NAOs. However in <u>Saccharomyces</u>, suitable microtubules do not appear to exist, hence a membrane associated process is favoured (Peterson & Ris, 1976). Recently, experimental evidence against a role for microtubules has apparently been obtained in <u>Saprolegnia</u> (Heath, unpublished and Fig. 14) where some centriole separation occurs in the presence of oncodazole (see Section VI), a drug which causes the loss of many of the cytoplasmic microtubules. However, since separation is not substantial nor are <u>all</u> cytoplasmic microtubules absent this data is not conclusive. In those organisms in which centriole migration occurs prior to spindle formation the problem is no less uncertain. For example, Kazama (1974) and McNitt (1973) have demonstrated centriole associated cytoplasmic microtubules which form and migrate with the centrioles. The formation of such microtubules must have functional significance yet the geometry of the nucleus precludes their functioning in a simple pushing manner by elongation. They are also associated laterally with the nuclear envelope thus further complicating the issue by allowing the hypothesis that the microtubules actively slide along the nuclear envelope dragging their centrioles with them. Furthermore, as noted in Section

III, nuclear envelope mediated events need not be synonymous
with membrane phenomena; there may well be a force generating
network associated with the nuclear envelope. That such a
system may exist in other organisms is shown by Kubai (1973)
for mitotic chromosome movements in Trichonympha and by the
behaviour of meiotic prophase chromosomes (reviewed by Moens,
1973). Clearly further analysis of the nuclear envelope could
be highly profitable and is needed before the mechanism of NAO
separation can be resolved. It must also be conceded that
there could be more than one mechanism operating in different
species.

One additional point concerning spindle formation should
be mentioned. It was assumed earlier that kinetochores and
chromosomes are associated with one another during early prophase. This assumption is valid for organisms such as Dictyostelium whose chromosomes are visible in the electron microscope (Moens, 1976) but may not be valid when the chromosomes
are not visible as, for example, in Saprolegnia and Saccharomyces. Byers and Goetsch (1975a) and Zickler and Olson
(1975) have shown that at meiotic prophase in Saccharomyces,
spindle microtubules and synaptonemal complexes coexist in the
nucleus. The complexes were in no way clustered around the
developing spindles thus suggesting a lack of association between the synapsed chromosomes and the spindle microtubules.
If the chromosomal microtubules are present at this time then
it seems that they do indeed form from the NAOs and subsequently associate with the chromatin. Because a) meiotic prophase lasts such a long time relative to mitotic prophase, b)
some of the strains used are mutants with defects in some
genes associated with meiosis and c) it is not known if the
chromosomal microtubules are present, an unequivocal conclusion about mitosis is not possible but clearly further work to
resolve this question is highly desirable.

The final aspect of spindle formation which requires
attention is the role and nature of the NAO associated material inside the nuclear envelope of a number of species. For
example in many Basidiomycetes (Girbardt, 1968, 1971, Raju &
Lu, 1973, Gull & Newsam, 1975, Harder, 1976a, Heath & Heath,
1976 and Fig. 7) and some Ascomycetes and Deuteromycetes
(McKeen, 1972, Aist & Williams, 1972) the region of the nuclear envelope adjacent to the interphase NAO is subtended by a
small, often hemispherical, region of amorphous material which
usually differs in its staining properties and general

appearance from the rest of the nucleoplasm and from the
chromatin (contrary to the labels in Girbardt, 1968, 1971) and
nucleolus. This material can no longer be recognized as the
spindle develops suggesting that it may be inter-converted into some component of the spindle. Its composition and function are totally obscure but should be born in mind in future
work on mitosis.

B. Structure

The structure of fungal spindles is of course diverse as
might be expected for a polyphyletic group of organisms. A
general synopsis is further complicated by tremendous heterogeneity in the level of analysis between different papers.
However a number of important points can be made.

1. Chromosomal Microtubules

There are no organisms in which a lack of chromosomal
microtubules has been convincingly demonstrated; conversely
many diverse species have been shown to possess unambiguous
chromosomal microtubules terminating in variously complex
kinetochores (Table 1). The only major group of fungi in
which there remains significant doubt is the zygomycetes in
which, excluding the highly atypical Basidiobolus, kinetochores and chromosomal microtubules have not been convincingly
demonstrated. The micrographs of Ancylistes (Moorman, 1976)
suggest, but do not prove, that chromosomal microtubules may
be present. Likewise pictures of Conidiobolus (Robinow, personal communication and Fig. 15) strongly suggests the presence of chromosomal microtubules and Figs. 5 and 6 of McCully
and Robinow (1973) are tantalizingly suggestive. However, it
is clear that in the later stages of mitosis shown by Franke
and Reau (1973) and McCully and Robinow (1973) there are no
chromosomal microtubules present. Their absence could be due
to differential lability of microtubules during fixation but
most data on mitosis indicates that at least in higher organisms, it is the non-chromosomal microtubules which are most
labile (e.g. Brinkley, et al., 1967, Brinkley & Cartwright,
1970). An alternative explanation favoured by this reviewer
is that chromosomal microtubules are present at the early
stages of mitosis but that they shorten and do their job

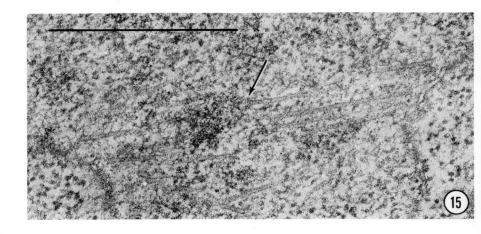

Fig. 15. <u>Conidiobolus</u> <u>villosus</u>. Non-median longitudinal section of a short spindle probably at metaphase. Note probable kinetochore (arrow) adjacent to the darker material which is presumably a chromosome. Bar = 1 μm. Unpublished, courtesy of Dr. C.F. Robinow, University of Western Ontario.

before the stages of mitosis detected by Franke and Reau (1973) and McCully and Robinow (1973) (excluding Fig. 5 in the latter paper). Clearly a more detailed study of the Mucorales is needed before the absence of chromosomal microtubules can be accepted as fact.

Among those fungi which have been studied in sufficient detail, it is evident that very few microtubules per chromosome is the rule (Table 1). One is the most common arrangement but up to five or more are rarely reported (Table 1). This is remarkably low when compared with the figures reported for higher organisms, for example 90-145 in <u>Haemanthus</u> (Jensen & Bajer, 1973). It is clear that the chromosomes of <u>Haemanthus</u> are much larger than those of the fungi thus there could well be a correlation between chromosome size and number of chromosomal microtubules attached to the chromosome. Unfortunately there is insufficient data on both chromosome size and microtubule number to permit an analysis of this possible correlation.

The structure of the point at which chromosomal microtubules attach to the chromosome is usually defined as the

kinetochore. Fungal kinetochores are typically small and
simple in morphology so that they are easily missed in material which is not optimally fixed and serially sectioned. In
those organisms listed as having only a single microtubule per
kinetochore, the kinetochore may be a disc which is not much
larger than the microtubule itself (e.g. Heath & Greenwood,
1968, 1970, Heath, 1974b, and Figs. 11-13). Reports claiming
the absence of kinetochores should be carefully evaluated for
their ability to detect such small structures. In this reviewer's opinion there is no compelling evidence to believe
that kinetochores are absent in any fungus, which is not of
course to say that they do occur in all species, but only to
date there are no convincingly documented absences. Although
the majority of fungal kinetochores are very simple, in Dictyostelium Moens (1976) has shown them to be quite large and
well differentiated. This occurrence is interesting in that
Polysphondelium, which is presumably closely related, has very
simple kinetochores (Roos, 1975). This variability suggests
that a) kinetochore morphology has little phylogenetic value
and b) the structural variations are probably not functionally very important since, in other aspects, the spindles of
these two cellular slime molds are similar and presumably
function in a similar manner.
     It has been possible to show that in many fungi the
chromosomal microtubules do indeed run the whole distance from
the chromosomes to the spindle poles. This is shown particularly convincingly in the reports of Lu (1967), Heath (1974b),
Peterson and Ris (1976), Roos (1975), Beckett and Crawford
(1970), Moens (1976), Aist and Williams (1972) and others and
is contrary to the report of Fuge (1974) for crane fly spindles. Such demonstrations are obviously important in understanding possible models of spindle mechanics. As far as can
be judged, this connection to the spindle pole, once established in prophase, is maintained throughout mitosis. This
means that there is considerable variability in length of
chromosomal microtubules in some species because, as shown in
Table 1, there is frequently an absence of a metaphase plate.
In species which lack a metaphase plate, the chromosomes are
dispersed, to various extents, along the length of the
spindle (e.g. Aist & Williams, 1972, Heath, 1974b, Heath &
Heath, 1976, Huang, et al., 1975, and other references in
Table 1 plus Figs. 16-19). This means that if the kinetochores are paired (as defined earlier), as they frequently are

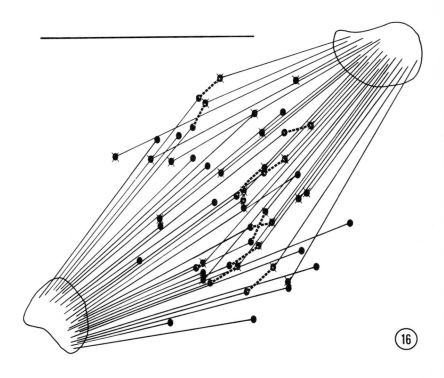

Fig. 16. <u>Saprolegnia ferax</u>. Diagrammatic isometric projection of a metaphase spindle showing the distribution of the kinetochores (solid 0) and the polar pocket regions of the nuclear envelope (irregular outlines). Kinetochores joined by dashed lines were judged to be paired with one another. Reconstructed from serial sections of the spindle shown in Fig. 17. Bar = 1 µm. Unpublished.

Fig. 17. <u>Saprolegnia ferax</u>. Median longitudinal section of a metaphase spindle showing pairs of centrioles at each pole and numerous kinetochores (e.g. arrows). Bar = 1 µm. Unpublished.

prior to anaphase, the chromosomal microtubules to opposite poles must vary considerably in length. This dispersion of the chromosomes along the spindle is responsible for the observations of the so-called "two-track" system of fungal mitosis, a topic which has been ably discussed by Aist and Williams (1972) and Robinow and Caten (1969). The important unsolved feature about the lack of a metaphase plate is its stability. In higher organisms it is well known that chromosomes oscillate considerably prior to the establishment of the metaphase plate (e.g. Nicklas, 1971, Rickards, 1975). This could occur in the fungi. Thus the explanation for the lack of a metaphase plate could simply be that the signal for anaphase chromosome separation is given before pre-metaphase oscillations have terminated, an entirely possible situation in organisms with a very fast mitotic system (see Section V D). Such an explanation would indicate that a metaphase plate is not a prerequisite for mitosis. Alternatively, the observed pre-anaphase distribution of chromosomes may be a static situation, as appears to be the case in <u>Fusarium</u> (Aist, 1969, Aist & Williams, 1972), which would mean that an equilibrium of equally opposed forces to establish a metaphase condition is unnecessary. Whatever the explanation, the result will be a significant contribution to our knowledge of the mechanics of mitosis in general.

After metaphase, with few exceptions (Table 1), the chromosomal microtubules shorten substantially so that by telophase they are very short and located at the spindle poles prior to their disappearance by interphase (e.g. Heath, 1976b, Heath & Greenwood, 1970, and Fig. 20). The exceptions listed in Table 1 illustrate the points that a) not all components of

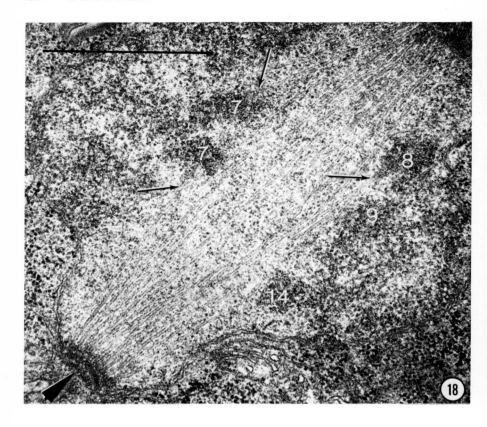

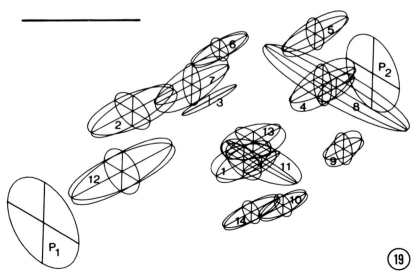

Fig. 18. <u>Uromyces phaseoli</u> var. <u>vignae</u>. Median longitudinal section of a metaphase spindle showing chromosomes (numbers correspond to those in Fig. 19), kinetochore microtubules (arrows) and one NAO inserted in the nuclear envelope (large arrow). Bar = 1 μm. From Heath and Heath (1976), courtesy of Rockefeller University Press.

Fig. 19. <u>Uromyces phaseoli</u> var. <u>vignae</u>. Isometric projection of a metaphase spindle derived from serial sections of the spindle shown in Fig. 18. The position of 14 chromosomes and the NAOs (P_1 and P_2) are shown. Bar = 1 μm. From Heath and Heath (1976), courtesy of Rockefeller University Press.

a mitotic system need to function in all organisms and b) anaphase movements have two components as indicated by Oppenheim, et al. (1973) for mammalian cells and Kerr (1967) for Myxomycetes.

2. Non Chromosomal Microtubules

It is clear that in a number of fungi the spindles contain continuous, interdigitating, free and polar microtubules (e.g. Heath, 1974b, Heath & Heath, 1976). What is not clear is the relative numbers of these tubules and the in vivo reality of such tubules. The reason for this uncertainty is of course entirely technical. It is difficult to unambiguously track microtubules over long distances yet such tracking is the only way to prove which category a particular microtubule belongs to. The least ambiguous method is serial cross section analysis but this requires long series, especially at anaphase and telophase, and very good fixation. The alternative is to use thick sections and high voltage microscopy but when the microtubules are closely spaced this is not unambiguous (Peterson & Ris, 1976). However a more serious limitation is the problem of preservation of the microtubules. Microtubules, especially those of the spindle, are known to be labile (e.g. reviews of Hepler & Palevitz, 1974, Snyder & McIntosh, 1976) thus they could easily partially depolymerize during fixation thereby converting continuous microtubules to any of the other categories of non-chromosomal microtubules. Evidence to suggest that this happens is shown by Luftig, et al. (1977) who demonstrated more, and longer, microtubules after fixation in microtubule polymerization permitting media.

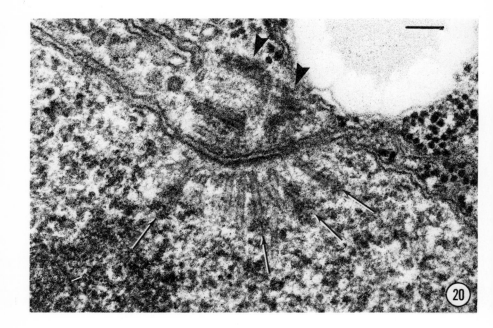

Fig. 20. Saprolegnia ferax. Portion of the polar region
of a telophase spindle showing kinetochores (arrows) close to
the nuclear envelope adjacent to the centrioles (large
arrows). Bar = 0.1 μm. Unpublished.

However since they had to incorporate guanosine triphosphate
in the medium to obtain this result, they could not rule out
artifactual polymerization of microtubules. At present there
is no alternative to electron microscopy to show the lengths
and numbers of individual microtubules so that an unequivocal
answer to the question of the degree of homology between the
fixed image and the in vivo spindle is impossible. With the
above cautionary remarks in mind, it is possible to continue
discussion of the non-chromosomal microtubules.

There appears to be some variability between different
organisms in the relative abundance of continuous versus interdigitating microtubules. For example, in Thraustotheca
(Heath, 1974b) and Saprolegnia (Figs. 21 and 22) continuous
microtubules are sparse and, at least at anaphase/telophase,
Moens (1976) suggests that all of the microtubules are interdigitating in Dictyostelium. However in Phycomyces (Franke &

Reau, 1973) and possibly in <u>Saccharomyces</u> (Peterson & Ris, 1976, Moor, 1966, Matile, et al., 1969) most of the non-chromosomal microtubules appear to be continuous and in <u>Uromyces</u> (Heath & Heath, 1976) continuous microtubules appear to be abundant throughout mitosis. The published data on other spindles is too ambiguous or insufficient to differentiate between the two types of microtubules. However the two different distributions (i.e. predominantly interdigitating <u>versus</u> predominantly continuous) have potentially major implications for the allowable mechanisms for spindle elongation. Obviously continuous microtubules alone <u>could</u> bring about spindle elongation by simple polymerization (although force generation by sliding is not precluded, see Section V E) but interdigitating ones could not do so since there is no obvious structure against which they can push on their free ends. That they do elongate during anaphase/telophase is not questioned (Heath, 1974b and Figs. 13, 21, 22), but if this elongation is force generating it implies that the nucleoplasm has structural properties against which they can push, a point discussed in more detail by Heath (1975) and in Sections V E and VI.

A further point of variability among the fungi is the arrangement of the non-chromosomal microtubules. In many species they are arranged in a tightly packed bundle in the centre of the spindle with the chromosomal microtubules arranged around their periphery (Table 1). In some cases the bundle appears to become more tightly packed during anaphase/telophase (e.g. Aist & Williams, 1972). In other species the non-chromosomal microtubules form a much more loose arrangement throughout mitosis (Table 1). There does tend to be some gradation in this pattern, especially in the homobasidiomycetes, but the majority of spindles fall clearly into one category or the other. It is possible that such variations are a fixation induced artefact but they seem to be too consistent in diverse fixations to be accounted for entirely in this way. The significance of these arrangements will again be discussed in Section V E.

The polar and free microtubules are not widely reported but the free ones at least have been seen in an Oomycete and a Basidiomycete (Heath, 1974b, Heath & Heath, 1976). No other reports are known in which they have been looked for but they do also occur in the spindles of higher organisms (Fuge, 1974, McIntosh, et al., 1975). Their function, if any, is totally obscure.

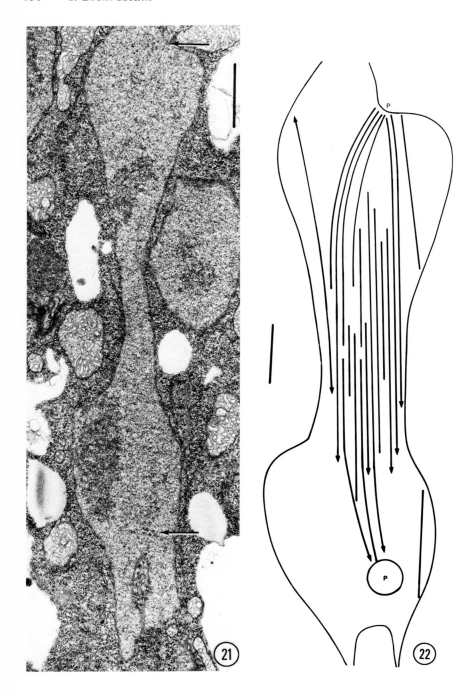

Fig.21. Saprolegnia ferax. Median longitudinal section
of a telophase nucleus. Spindle poles are indicated by arrows. The nucleolus and nuclear envelope are clearly persistent. The nucleoplasm extends well beyond the spindle poles.
Such extensions were associated with cytoplasmic microtubules.
Spindle microtubules are barely discernible at this magnification. Bar = 1 μm. Unpublished.

Fig. 22. Saprolegnia ferax. Reconstruction (laterally
expanded) of part of the spindle shown in Fig. 21. Few microtubules extend from pole to pole. Those ending in arrowheads
continued for an undetermined distance beyond the end of the
series of sections available. Bar = 1 μm. Unpublished.

3. Non Microtubule Components

a. Chromosomes. The structure and behaviour of fungal
chromosomes has been a topic of much concern to cytologists
for many years. The reason for this concern has primarily
been the difficulty of obtaining mitotic figures comparable to
those obtainable from higher organisms with classical light
microscope staining techniques. The use of the electron microscope has not improved the situation, at least in some respects. Whilst in many fungi, especially the Basidiomycetes
(e.g. Coprinus [Lu, 1967], Uromyces [Heath & Heath, 1976],
Poria [Setliff, et al., 1974]), higher Ascomycetes (e.g.
Ascobolus [Wells, 1970, Zickler, 1970]) and myxomycetes (e.g.
Acryria [Mims, 1972], Physarum [Aldrich, 1969], Echinostelium
[Haskins, 1976])well stained and discrete chromosomes are present in thin sectioned material, in other species it is very
difficult to detect chromosomes at any stage of mitosis (e.g.
Thraustotheca [Heath, 1974b], Saccharomyces [Peterson & Ris,
1976]) even when using techniques (Aist & Williams, 1972)
which enhance the appearance of hard-to-see chromosomes in
other fungi (Heath, unpublished observation). The reasons for
these problems are obscure. It has been said many times before that fungal chromosomes are very small and this is unden-
iably true. For example it can be calculated that each chromosome of Physarum contains approximately an order of magnitude
less DNA than the average mammalian chromosome (Mohberg, 1977,
Shapiro, 1976). This difference is accompanied by a comparable size difference whereby the chromosomes of Echinostelium
(Haskins, 1976) for example are less than 1 μm long whereas

those of mammals range from 1 to 10 μm (Arrighi, 1974). However size differences should not be a problem at the electron microscope level of resolution, a point which is clearly made by the highly visible myxomycete chromosomes seen in any of the references in Table 1. Likewise, for the same reason a small DNA content cannot explain the poor visibility of fungal chromosomes in general since there is a roughly constant amount of DNA per chromosome in a number of species including Physarum yet some are clearly visible and others are not.

Lack of visibility of fungal chromosomes may be due to differences in chromosomal proteins relative to higher organisms. It has been said that fungi lack the most well known chromosomal proteins, the histones (Leighton, et al., 1971) but more recent studies indicate that some histone like proteins are present in a number of fungi such as Neurospora (Hsiang & Cole, 1973), Achlya (Horgen, et al., 1973) and Saccharomyces (Franco, et al., 1974, Tonino & Rozijn, 1966, Wintersberger, et al., 1973) and that the full complement as found in calf thymus is present in Aspergillus (Felden, et al., 1976). The ratio of histone to DNA is perhaps lower in some fungi (Hsiang & Cole, 1973) but this could be due to technical problems (Felden, et al., 1976). In any event, in all of the above species the contrast of the chromosomes in electron microscopy sections is poor, yet because all have at least some histones, that absence of histone alone cannot be the reason.

A very probable cause of difficulty in discerning fungal chromosomes is the presence of a lower degree of condensation during mitosis when compared with other organisms. Data on this point is sparse. Peterson and Ris (1976) suggest that condensation does not occur in Saccharomyces, a point supported by the use of different, although not entirely convincing, techniques by Gordon (1977) in the same species. However Wintersberger, et al. (1975) show structures interpreted as condensed chromosomes in Saccharomyces. They obtained a mean of n = 18 which is close to the value of n = 17 obtained by linkage group analysis for this species (Mortimer & Hawthorne, 1969) but their wide spread about this mean does not enhance one's confidence in the ability of the technique to reveal the in vivo state of the chromosomes; the possibility of artifactual condensation induced by their processing seems very real. A similar analysis by Fischer, et al. (1975) on Schizosaccharomyces gave a value of n = 8 which is at variance

with the genetic linkage group (Kohli et al., 1977) and light
microscopy (Robinow, 1977) value of n = 3 for the same species.   Thus the degree of condensation of fungal chromosomes
relative to other organisms remains an open question as does
the reason for their indistinct appearance in thin sections of
many species.

The final point concerning fungal chromosomes is their
anomalous arrangement in some species. The absence of a metaphase plate has already been discussed but, at least in many
of these species, discrete chromosomes are present as mentioned above. However in a number of species, notably in the
Plasmodiophoromycetes, Chytridiomycetes and Hyphochytridiomycetes (references in Table 1) the converse situation occurs.
The chromosomes are found at metaphase as a very dense plate
with no evident separation into separate chromosomes. This
dense packing persists through anaphase when two very distinct flat plates move apart prior to rounding up during telophase. The plates have numerous kinetochores on them and are
penetrated by non-chromosomal microtubules. How this behaviour relates to the packaging of the DNA is totally obscure
but, from the point of view of the present review, it is quite
evident that a plate of chromatin must present a very different mechanical proposition when compared with individual
chromosomes. Whether there are any fundamental differences in
mechanisms of force generation during mitosis remains to be
seen, but this plate like arrangement is not unique to the
fungi; a similar arrangement is found in a number of algae
(e.g. Oakley & Dodge, 1976). It must of course be remembered
that the plate like array could also be an artefact of fixation. However, if so it must still reflect some differences
in chromatin organization since it is not universal, discrete
chromosomes may also be seen in many spindles of diverse organisms.

b. Filaments. It would not be an understatement to say
that the reports of filaments in fungal spindles are sparse.
The only report known to this reviewer is that of Ryser (1970)
who showed numerous 3-9 nm diameter filaments in the plasmodial mitotic spindles of Physarum. These filaments were not
characterized beyond their not being breakdown products of
microtubules, nor was their distribution given in detail.
However the lack of reports of filaments in spindles is not in
any way a good indication of their absence. In many reports

of fungal spindles the quality of fixation is too poor to reveal such structures which, in other organisms, are known to
be more fixation labile than microtubules (Tilney, 1976). It
should also be remembered that there are reports of filaments
of unidentified composition in the nucleoplasm (see Section
VII). Some of these could be actin which could then be potentially available for the spindle. In view of the data reviewed by Forer in this volume, the search for filaments in fungal
spindles, specifically actin filaments, should be pursued
strongly in the fungi as well as elsewhere. However, as mentioned above, the results of such a search should be interpreted with extreme caution; negative results would be very
hard to prove unambiguously.

c. Vesicles. A number of workers (Harris, 1975; Wicks &
Hepler, 1976) have recently taken the observations of Weisenberg (1972) on the effect of calcium ions on microtubule polymerization as an indication of a possible mitotic spindle control system, using the analogy of the sarcoplasmic reticulum
calcium pump in muscle movements to support this possibility.
It is in this context that the vesicles adjacent to the chromatin masses in Sorosphaera (Braselton, et al., 1975) and Rhizidiomyces (Fuller, personal communication) are of great interest but little proven significance. Their presence may indicate some role in ion control within the spindle. Conversely
they may merely represent a variation on the theme of daughter
nuclear envelope formation, perhaps being a step further along
a line shown by Entophlyctis (Powell, 1975). Their role is
clearly unknown but their existence should be familiar to
those with an interest in the mechanisms of mitosis.

C. Composition

Data on fungal spindle composition are almost non-exist-
ent. There have been no spindle isolation experiments of the
type performed on higher organisms so that even this crude
means of analysis has not been possible in the fungi. However
some deductions are possible. Clearly tubulin is present.
The homology of this tubulin to that of other organisms is
obscure. Fungi are well known to be highly resistant to the
action of the tubulin binding drug, colchicine (see Section
VI) and Burns (1973) has suggested that, whilst a tubulin-like
protein is present in Schizosaccharomyces, it does not bind

colchicine. In contrast to this report, Haber, et al. (1972),
Olson (1973), Heath (1975) and Davidse and Flach (1977) have
described colchicine binding components in a variety of fungi.
However the characteristics of some of these components indicate that, if they are indeed tubulin, the tubulin may not be
identical with that of mammalian brain. For example they
differ from brain tubulin in having a much greater affinity
for MBC (Section VI C)(Davidse & Flach, 1977). In contrast
Sheir-Neiss, et al. (1976) and Davidse and Flach (1977) have
demonstrated tubulin like proteins in Aspergillus which are
electrophoretically the same as those of mammalian brain.
These proteins co-polymerize with brain tubulin (Davidse,
1975, Sheir-Neiss, et al., 1976). Water and Kleinsmith (1976)
have also shown similar proteins in Saccharomyces. Jockusch
(1973), on the other hand, reported the absence of any tubulin-like molecules in Physarum. This absence could easily be
due to technical problems of extraction, or the activity of an
endogenous protease. However, it is evident that the degree
of homology between various fungal tubulins, and the tubulins
of other organisms, is still an open question. Furthermore,
variability in the tubulins among a polyphyletic group such as
the fungi is very likely to occur, especially since the tubulins in different strains of a single species appear to vary
somewhat in their properties (Davidse & Flach, 1977).

The identity of other spindle components is totally obscure. The filaments reported by Ryser (1970) could be actin,
but there is no proof of this. Likewise Jockusch and her co-
workers (Jockusch, et al., 1973, 1974, Hauser, et al., 1975)
have presented evidence for the presence of actin and myosin
in interphase Physarum nuclei but whether or not any of these
molecules are a component of the spindle is obscure. It has
been suggested that Physarum nuclear actin and myosin are involved in chromosome condensation (Lestourgeon, et al., 1975).
Since microtubules are likely to comprise as little as 1% of
the mitotic spindle (Forer, 1974), there is clearly much to be
accounted for. Apart from the search for contractile proteins
it might be profitable to also search for ATPases of the type
described by Petzelt and Auel (1977).

D. Speed and Efficiency

Data on the time taken for mitosis in a variety of fungi
is shown in Table II (see also Heath & Heath, 1978). As

TABLE II   Mitotic Times for Diverse Fungi

Genus	Reference	Time (min.)				
		Prophase	Metaphase	Anaphase	Telophase	Total
Physarum (am. & plas.)	Koevenig, 1964	15-60	1-5	1-5	2-5	avg.30
Didymium (amoebae)	Kerr, 1967	2	3	1-2	1	7-8
" (plasmodia)	" "	3-5	6-7	1	1	10-12
Perichaena (amoebae)	Ross, 1967	2-3	1-2	<1	3	7-8
Allomyces	Olson, 1974b					5 ?
Mucor	Robinow, 1962, 1957				4 ?	100
Basidiobolus	Robinow, 1963					15-20
Strongwellsea	Humber, pers. comm.				30	240
Endogone	Bakerspigel, 1958					4-5

Genus	Reference						
Saccharomyces	Robinow & Marak, 1966						10-15
Schizosaccharomyces	McCully & Robinow, 1971						30 ?
Aspergillus	Robinow & Caten, 1969						6-8
Neurospora	Bakerspigel, 1959						10
Ascobolus	Zickler, 1971	2		1	1	3	6-7
Podospora	"	2		1	1	3	6-7
Fusarium	Aist & Williams, 1972	1.2		2	0.2	2.1	5.5
Ceratocystis	Aist, 1969						5-10
Rhodosporidium	McCully & Robinow, 1972a						< 15
Aessosporon	" "						< 15
Leucosporidium	" 1972b						< 15
Ustilago	Poon & Day, 1973	12		5	28		45
Polystictus	Girbardt, 1961		1.8		2	1.3	5-6
Various Homobasids.	Thielke, 1973						5-7

143

previously pointed out by Robinow and Caten (1969), these
times are in many cases open to doubt because one can only
measure the visible stage of mitosis, early prophase events
are likely to be undetected at the light microscope level.
However the clear impression is that the time taken for mitosis in the fungi is often very short when compared with the
times for higher organisms (see Mazia [1961] for a range of
values). The consequences of this rapidity as far as mitosis
is concerned are obscure, but it does make the hypothesis,
that the absence of a metaphase plate is due to a dynamic os-
cillation of chromosomes right up to anaphase, more tenable
(see Section V Bl). Another aspect of mitosis which may be
related to the speed of the system is the apparent lack of
efficiency in obtaining equitable genome distribution (Day,
1972). The data for this statement is in the observations of
the high frequencies of aneuploidy resulting from mitosis in
Aspergillus and Verticillium (Käfer, 1961, Hastie, 1962, Upshall, 1966, Faulkner, 1967). A shortage of comparable data
for other organisms makes it hard to do a meaningful analysis
of the relative efficiencies of various mitotic systems but
the above data does suggest caution in ascribing equivalent
functional ability to all variations of mitotic spindles en-
countered in "lower" organisms.

E. Mechanisms of Force Production?

    As indicated by Forer (this volume) the current state of
knowledge concerning force production during mitosis can be
summarized in two sentences. a) We have no unequivocal data
on the precise mechanisms of force generation in any mitotic
spindles. b) There are numerous extant hypotheses, each of
which has observations both for and against their various
features. However the data on fungal mitoses can provide some
useful indicators to possible mechanisms and can place restrictions on extant hypotheses. The first point which must be
made is whether or not fungi are likely to utilize fundamentally similar mitotic force generating mechanisms to those used
by "higher" plants and animals. No unequivocal answer can be
found to this question but one can certainly argue that since
conservatism of the basic process seems to be the rule for
other major cellular activities, such as flagellum axoneme
structure and function, pathways of photosynthesis, respira-

tion, transcription and translation, then it is likely that
similar conservatism will occur in the fundamental force generating aspects of mitosis. Only two features of fungal mitosis suggest major differences relative to "higher" organisms.
The early separation of chromosomes into distinct groups during prophase (see Section V A) appears to be unusual but mech-
anically it may be achieved by processes similar to those moving chromosomes or synaptonemal complexes during prophase
(e.g. Rickards, 1975). The other difference appears to be one
of speed. Anaphase chromosome movements in many mitotic systems are in the order of 1 μm/min (Mazia, 1961) but from the
data of Aist (1969) and Aist and Williams (1972) the velocities in Fusarium must be up to 9 μm/min, almost an order of
magnitude larger. It is unclear whether this difference is an
indication of a fundamentally different force generating system or not. However Forer (1974) has argued that chromosomal
microtubules act as dampers of the anaphase force generating
system. Since fungi typically have fewer chromosomal microtubules than other organisms (Section V B1) the higher anaphase speeds may simply be a response to fewer "dampers".
Thus these two seemingly anomalous points can easily be accom-
modated in the general data applicable to "higher" organisms.
As noted in the preceding sub-sections, there is little other
reason to believe that the fungi have "abnormal" mitotic mechanisms. What then can the fungi contribute to general theories of mitotic mechanisms?

Fungi with close packed truly continuous microtubules
could generate force for spindle elongation by microtubule
polymerization (Inoué & Sato, 1967). Conversely force could
be generated by inter-microtubule cross-bridge mediated active
sliding of the type envisaged by McIntosh, et al. (1969), with
concomitant passive microtubule polymerization. However, widely spaced, interdigitating microtubules of the Thraustotheca
type spindle (Heath, 1974b) must interact in some way with the
nucleoplasm if they are to generate force since they can hardly slide relative to each other nor push against the opposite
spindle pole (Heath, 1975a). Microtubule interaction with a
gel like spindle matrix (nucleoplasm) has been postulated by
Dietz (1972).

The present absence of actin data for fungal spindles
(Section V B 3b) means that fungal work cannot contribute,
either positively or negatively, to the arguments for action
involving hypotheses (Forer, 1974 and this volume). Movement

generated by the commonly observed single kinetochore microtubules (Section V B1) cannot be explained by the zipper hypothesis (Bajer, 1973) as stated, but various modifications of
that hypothesis can not be formally ruled out by any fungal
data. As emphasized earlier (Section V B1), if the non-plate
metaphase conditions are unambiguously static, then positioning of the chromosomes within the spindle need not involve a
balance between equally opposed active forces since such are
unlikely to exist if the forces are proportional to length of
chromosomal microtubules or chromosome to pole distances as is
often suggested (McIntosh, et al., 1969, Subirana, 1968, Östergren, 1950, see also discussions in Bajer & Mole-Bajer,
1972). For a more extended discussion along these lines the
reader is referred to Heath (1975a).

The above points illustrate that at least some aspects of
fungal mitoses can place restrictions on universal hypotheses
for mitosis and thus serve to emphasize the point that further work on these organisms may have broad significance to
studies of mitosis in "higher" organisms. The other obvious
point is that prior to work on fungal mitotic systems an investigator should become familiar with the current, and older,
literature from other kingdoms of organisms.

VI. ANTIMITOTIC AGENTS

In order to understand how the various components of a
mitotic system contribute to the overall process it would be
very helpful to have agents which could specifically and re-
versibly block the activity of each component. Unfortunately
at present the range of agents available is restricted primarily to antimicrotubule compounds and even they are not as well
characterized as would be most desirable. The two major problems with available antimitotic agents are, a) a lack of
carefully demonstrated specificity and b) an absence of agents
with known activity against a number of the components which
may be involved in mitosis. For example there are no known
agents which specifically block the activities of actin (with
the possible exception of phalloidin [Dancker, et al., 1975]),
myosin, dynein, various known ATPases, kinetochores and NAOs.
Those compounds with some demonstrated, or potential, activity
against fungal mitosis will be discussed below.

A. Colchicine and Colcemid

Colchicine, and its derivative, colcemid, are the best known antimitotic agents. In many cells they have been shown to block mitosis at metaphase and lead to polyploidy by preventing the formation of microtubules (or causing dissolution of previously formed tubules) by binding to the tubulin dimer subunits of microtubules. Each dimer has one colchicine or colcemid binding site (see reviews of Margulis, 1973, Olmsted & Borisy, 1973, Hepler & Palevitz, 1974). Whilst the tubulin dimer is often seen to be the main binding component in cell extracts, there are numerous reports showing that these alkaloids inhibit DNA, RNA and protein synthesis (Kuzmich & Zimmerman, 1972 and references therein), bind to membranes (Hotta & Shephard, 1973, Stadler & Franke, 1974, Bhattacharyya & Wolff, 1975) and generally cause various non specific cellular damage, including death, when used at high concentrations (Eigsti & Dustin, 1955, Heath, 1975b). These reports emphasize the lack of specificity of these drugs which are perhaps the most intensively studied antimitotic agents. This point is especially important in studies on fungi where millimolar concentrations are needed to obtain an effect comparable to that caused by micromolar concentrations in animal cells.

To date the use of colchicine and colcemid on fungi has not been extensive nor very productive in explaining how mitosis works. Eigsti and Dustin (1955) summarized the early literature on the use of colchicine on fungi and concluded "That mycelial growth may be influenced is probable but polyploidy or induction of mutations is extremely doubtful." More recent work has not clarified the response of fungi to these drugs. Haber, et al. (1972) and Lederberg and Stetten (1970) showed that colcemid could inhibit cell division in Saccharo-myces and Schizosaccharomyces respectively but this effect required $1-10 \times 10^{-3}$ M concentration and was dependent on the phase of growth of the culture and the composition of the medium (Haber, et al., 1972). In Basidiobolus, Gull and Trinci (1974) were unable to block hyphal mitosis with 1.25×10^{-2} M colchicine and Davidse and Flach (1977) found that 10^{-2} M colchicine did not inhibit mycelial growth in Aspergillus. Girbardt (1962) noted that up to 7.5×10^{-2} M colchicine had no effect on vegetative nuclei of Polystictus. Likewise Slifkin (1967) found that Saprolegnia hyphal mitosis resisted 2.5×10^{-3} M colchicine and Heath (1975b) reported apparently

TABLE III   Variation in kinetochore numbers in <u>Saprolegnia</u>
nuclei exposed to various anti-mitotic agents

Treatment	Number of spindles analysed	Mean number of kinetochores (± standard deviation)	Value of p relative to control
Untreated (control)	23	17.0 ± 4.1	-
Cold (3°C, 1h)	6	17.8 ± 3.2	ns
Colchicine (10 mM, 24h)	13	32.2 ± 5.0	< .001
Colcemid (5 mM, ¼ h and 2 mM, 24h)	10	27.8 ± 2.2	< .001
Camphor (0.66 mM, 24h)	6	27.6 ± 3.1	< .005

normal spindles and microtubules after 24 h in 10^{-2} M colchicine and 5 x 10^{-3} M colcemid in the same genus. However in
this latter work hyphal growth was reduced and in 5 x 10^{-3} M
colcemid many hyphae were dead. More recently both colchicine
and colcemid have been found to give apparent tetraploids in
vegetative <u>Saprolegnia</u> hyphae where normal looking spindles
are found (<u>Heath, 1977</u> and Table III). That the response of

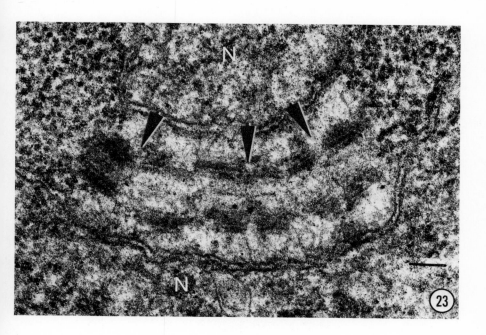

Fig. 23. <u>Saprolegnia</u> <u>ferax</u>. Three pairs of centrioles
(arrows) adjacent to a nucleus (N) in a sporangium induced to
differentiate in 7.5 x 10⁻³ M colchicine. Bar = 0.1 μm.
Unpublished.

this genus, and possibly other fungi, to these drugs is complex, and may depend on external and internal cellular envir-
onments, is shown by Slifkin's (1967) observation of colchicine blockage of oogonial nuclear divisions and the comparable
findings of abnormal zoospore formation (a microtubule involving process) (Slifkin, 1967) and multiple centrioles (Fig. 23)
in developing sporangia. Furthermore, some germinating zoospores contain fibrils (which may be colchicine induced microtubule degradation products) but no microtubules only after
colchicine treatment (Heath, et al., 1970, Heath, 1969). A
similar variation in cell response to colchicine depending on
cell physiology or culture conditions is found in <u>Allomyces</u>
where Olson (1972) has presented data suggesting that germinating spores become decreasingly sensitive to colchicine induction of metaphase arrest. Unfortunately the presented data

does not rule out the alternative simple hypothesis of colchicine-induced delay of mitosis, a hypothesis which would be compatible with the growth rate reduction noted above in Saprolegnia hyphae.

It is, perhaps, not unfair to summarize that colchicine and, or, colcemid may have some antimitotic effect on some fungi in some growth phases but often only at high concentrations, and more often they have little or no antimitotic activity. Thus these drugs are perhaps not the ones of first choice for use on fungi. However because they do have some activity in some stages they should be included in any initial investigation of microtubule dependent processes.

The basis of the lack of effect of colchicine and colcemid on some fungi is essentially unknown. A number of possible reasons, including the obvious permeability problem, have been discussed and discarded by Haber, et al. (1972) and Heath (1975b). As noted earlier (Section V c) there is some reason to believe that there are colchicine and colcemid binding components in various fungi (Haber, et al., 1972, Olson, 1973, Heath, 1975c, Davidse & Flach, 1977). Whilst some of these components are probably tubulin (Haber, et al., 1972, Davidse & Flach, 1977) it has been pointed out (Davidse & Flach, 1977) that the trichloracetic acid stability of some of the binding components described by Olson (1973) and Heath (1975c) is not what would be expected from a normal tubulin-colchicine complex. Furthermore, the possibility of radioactive contaminants in labelled colchicine solutions must be seriously considered when working with low levels of binding material such as are typically found in fungi (Davidse & Flach, 1977). However, it seems probable that, in at least some fungi, colchicine and, or, colcemid do bind to tubulin but probably with a lower affinity than that found for mammalian tubulin (Davidse & Flach, 1977, Haber, et al., 1972). Whether this difference is enough to account for the observed insensitivity in vivo is obscure. A comparison of the colcemid sensitive wild type and the colcemid insensitive mutants of Schizosaccharomyces (Lederberg & Stetten, 1970) would be most rewarding in clarifying the situation, as would investigations on the sensitive and insensitive life cycle stages of Saprolegnia as outlined above.

B. Griseofulvin

Griseofulvin is a Penicillium product which has been used
as an antidermatophyte drug. Details of its mode of action
and effects on diverse organisms have been ably reviewed by
Bent and Moore (1966) and need not be recapitulated here.
Two essential points emerge from their review. 1) Griseofulvin can have antimitotic colchicine-like effects on cells of
various organisms and (2) it has numerous effects on fungi, as
well as other organisms, which are unlikely to be explicable
on the basis of antimitotic or antimicrotubule activity.
Since the work of Bent and Moore (1966), a number of studies
have emphasized the point that griseofulvin can apparently re-
versibly block mitosis in such diverse fungi as Aspergillus
(Crackower, 1972), Basidiobolus (Gull & Trinci, 1973, 1974a),
Protomyces (Trinci & Gull, 1973) and Physarum (Gull & Trinci,
1974b) although it was ineffective in Saprolegnia (Heath,
1975b). This lack of effect in an Oomycete is not surprising
in view of the general lack of sensitivity of this group as
noted in Bent and Moore (1966). Recently Roobol, et al.(1976,
1977) have suggested that at least part of the antimitotic
activity of griseofulvin is due to its interaction with high
molecular weight, microtubule associated proteins (Snyder &
McIntosh, 1976) which are important in controlling microtubule
polymerization. This is the first report of an antimicrotubule agent which acts on a microtubule component other than
tubulin. The lack of effect on Oomycetes raises the interesting possibility that different organisms have different microtubule associated proteins which could then explain the differential stability of different microtubules to diverse dis-
rupting agents.
At present it is fair to conclude that griseofulvin does
interact with some microtubules and may block mitosis in some
fungi and thus has potential as an experimental tool. Unfortunately the lack of specificity mentioned above (point 2)
places considerable restrictions on this potential. It must
also be remembered that some fungi can transform griseofulvin
into inactive derivatives (Boothroyd, et al., 1961, Bod, et
al., 1973).

C. MBC

MBC (methyl benzimidazol - 2 - yl carbamate) is the

primary fungitoxic component of the systemic fungicide, benomyl. It disrupts mitosis in animal (Styles & Garner, 1974)
and plant cells (Richmond & Phillips, 1975) as well as in a
number of fungi such as Ustilago, Saccharomyces (Hammerschlag
& Sisler, 1973), Aspergillus (Davidse, 1973, Kunkel & Hädrich,
1977) and Botrytis (Richmond & Phillips, 1975). It also produces changes in hyphal organization which may be attributable
to antimicrotubular activity (Howard & Aist, 1977). Judging
by the list of fungi sensitive to benomyl (Edgington, et al.,
1971), it is probable that MBC blocks mitosis in a broad range
of fungi. However it does not affect all fungi; all oomycetes
and zygomycetes so far examined, and a number of basidiomycetes, are resistant (Edgington, et al., 1971, Staron & Allard,
1964, Bollen & Fuchs, 1970). It may be that some of the other
benzimidazole derivatives (Seiler, 1975, Davidse & Flach,
1977) could be antimitotic in MBC insensitive organisms since
oncodazole (see below) is effective in the oomycete Saprolegnia (Fig. 14 and Heath, unpublished).
      The mode of action of MBC has been substantially clari-
fied by Davidse (1975) and Davidse and Flach (1977) who show
that it binds to fungal tubulin, probably at the same site as
colchicine and colcemid. This suggests that it acts in an
analogous manner to colchicine but in vitro inhibition of
fungal tubulin polymerization has not yet been demonstrated
and the alternative experiment of demonstrating in vitro inhibition of mammalian tubulin is not possible since MBC has a
low affinity for mammalian tubulin. Variable affinity of tubulins from different species may well explain the differing
sensitivities of organisms to this compound, especially as MBC
resistant strains of Aspergillus have differing MBC-tubulin
binding constants (Davidse & Flach, 1977).
      The broad spectrum of fungal species sensitive to MBC
indicates that this agent has good potential as a tool for in-
vestigating fungal mitosis but again caution is necessary.
Insufficient biochemical work has been performed to adequately
describe possible non-microtubule based side effects. However, recently Bourgois, et al. (1977) have shown ultrastructural evidence for diverse side effects and Kumari, et al.
(1977) have shown that DNA synthesis can be inhibited inde-
pendently of mitosis; thus specificity for microtubules seems
to be lacking. Also some fungi are able to metabolize MBC to
less fungitoxic derivatives (Davidse, 1976).

D.   Oncodazole

Oncodazole is a benzimidazole derivative which has been developed as an antitumour drug by Janssen Pharmaceutica in Belgium.   It prevents in vitro polymerization of mammalian brain microtubules by binding to the colchicine binding site of tubulin dimers but it does not cause depolymerization of intact in vitro formed microtubules (Hoebeke, et al., 1976). As might be expected it therefore interferes with both cyto-plasmic and mitotic microtubules in living mammalian cells (De Brabander, et al., 1975, 1976). Davidse (1975) and Davidse and Flach (1977) have shown that it binds to fungal tubulin and inhibits growth of Aspergillus and Heath (unpublished and Fig. 14) has found that it blocks mitosis and causes a loss of cytoplasmic and non-chromosomal microtubules in Saprolegnia; thus it seems to have good potential for further use in fungi. Again the specificity of the compound is not adequately estab-lished but at least it is active in Saprolegnia at 5×10^{-6} M which is in a range where more selectivity and less general cytotoxicity is likely in comparison with the work on colchi-cine described above.

E.   Other Potential Antimitotic Agents

As shown in Table IV, there are a large number of other agents, both physical and chemical, which have been shown to have some antimitotic or antimicrotubule effects in some or-ganisms, including fungi. In general terms there has been too little work on the effect of these agents on fungi for worth-while generalizations to be made. For this reason the agents are listed in table form and references are made to papers in which their activities are established and their effects on fungi noted. Investigators interested in using any of these agents are advised to examine the relevant literature and break new ground! To date none of these compounds have made a significant impact on our knowledge of fungal mitosis but this may be more from lack of investigation than lack of pot-ential.

TABLE IV Potentially Useful Antimitotic Agents

Agent	General references to mode of action	Mode of action	References in which an antimitotic or antimicrotubule effect has been shown in fungi	References in which no specific antimitotic or anti-microtubule effect could be detected in fungi
Vinblastine or Vincristine	Wilson, et al.(1975)	Binds to Tubulin	Gull & Trinci (1974a)	Heath (1975b)
Podophyllin	Wilson, et al.(1975) Cortese, et al.(1977)	Binds to Tubulin	Ormerod, et al.(1977)?	Unknown
Camphor	?	?	Sansome & Harris (1962) Heath (1977)	Heath (1975b)
Melatonin	Winston, et al.(1974)	Binds to Tubulin	Unknown	Unknown
Veratrine	Witkus & Berger (1944)	?	Unknown	Heath (1975b)
Amiprophos-methyl	Kiermayer & Fedtke (1977)	?	Unknown	Unknown

154

Agent	Reference	Effect		
Isopropyl phenyl carbamate (IPC)	Coss, et al.(1975)	Anti MTOC	Gull & Trinci (1974a)	Unknown
Maytansine	Mandelbaum-Shavit, et al.(1976)	Binds to Tubulin	Unknown	Unknown
Trifluralin	Banerjee, et al.(1975)	?	Unknown	Unknown
Phalloidin	Dancker, et al.(1975)	Anti actin	Unknown	Unknown
High hydrostatic pressure	Zimmerman (1970)	Complex	Heath (1975b)	Heath (1975b)
Low temperature	Tilney & Porter (1965)	Complex	Unknown	Heath (1975b)

VII. NUCLEOPLASM

The nucleoplasm is traditionally perhaps the most neglected component of nuclei yet it must form a large portion of
the total volume of a nucleus. Even the definition of nucleoplasm is neglected and is of necessity a negative one; essentially everything inside of the nuclear envelope, which is not
otherwise identified, may be defined as the nucleoplasm. However, it is obvious that the composition of the nucleoplasm
must have some effect on mitosis if in no other way than its
being the medium in which the whole process occurs and through
which chromosomes move. This statement is especially relevant
to organisms, such as the majority of the fungi, in which the
nuclear envelope remains intact during mitosis. Unfortunately
we know very little about the nucleoplasm, either from fungi,
or other organisms but there are a few points which should be
made. It has been suggested that in a number of fungi the
spindle does not in fact provide the force for nuclear elongation because the nucleus is often elongated beyond the poles
of the spindle (Robinow & Marak, 1966, McCully & Robinow,
1971, 1973). In Saprolegnia and Thraustotheca such extensions
appear to be generated by cytoplasmic microtubules (Heath &
Greenwood, 1970, Heath, 1974b) but in Mucor such an arrangement was sought and specifically excluded (McCully & Robinow,
1973). McCully and Robinow (1973) argue that nuclear envelope
growth is responsible for nuclear elongation. Such could be
true but since the nucleoplasm changes shape concomitantly it
is equally probable that it is actively respnsible for nuclear
shape changes. Furthermore Aist and Williams (1972) have
pointed out a third possibility, that the condensed chromosomes at the spindle poles could generate the observed shapes.
There is no hard data to support any of these three contentions but the observations of nucleoplasmic actin and myosin
in Physarum (see Section V C) may indicate a contractile ability for at least part of the nucleoplasm. Likewise Berezney
and Coffey (1977) and Wunderlich and Herlan (1977) have recently isolated a largely proteinaceous nuclear matrix from
mammalian and Tetrahymena nuclei. Such a matrix is formally
part of the nucleoplasm and may have contractile properties
(Wunderlich & Herlan, 1977). However, it does not appear to
be an actin-myosin based system. Such results provide a very
timely reminder that the nucleoplasm is much more complex than
previously considered; thus considerations of mechanisms of

mitosis and nuclear shape changes should bear its potential contributions in mind.

The second point which should be made about the nucleoplasm is the fact that it is frequently reduced in volume at some point during mitosis. This is achieved, at least in part, by pre-prophase or telophase exclusion of sizeable portions of the nucleus, with or without concomitant nucleolus exclusion. Examples of organisms in which such processes occur are given in Table 1 (see also Savile, 1939, Craigie, 1959, Maheshwari, et al., 1967, Girbardt, 1968, Thielke, 1973, Harder, 1976a and b). Since the reasons for such a system are totally obscure, as are the mechanisms by which it might be achieved, further discussion is pointless. However it is a phenomenon which future investigators should look for and bear in mind when discussing the behaviour of nuclei and the nucleoplasm.

Another point which may be important is that, whatever the structure and function of the nucleoplasm, it does appear to undergo changes prior to, or during, mitosis. For example, Howard and Aist (1977) note that the nucleoplasm becomes less phase "dense" at the onset of mitosis. Similarly Robinow and Caten (1969), Poon and Day (1974) and Heath and Heath (1978) show changes in the contrast of the nucleoplasm during mitosis. The significance of these changes is unknown but they do emphasize the point that the nucleoplasm does change during mitosis. Such changes may or may not have functional significance but they must ultimately be explained.

The final point of information concerning the nucleoplasm is that it often contains fibrous material. Bundles of fibres, each fibre ranging from 2 to 15 nm in diameter, have been reported in variously treated and untreated nuclei of Neurospora (Beck, et al., 1970, Allen, et al., 1974) and some Oomycetes (Heath, et al., 1970, Gleason, 1973, Heath, 1975b, Ellzey, et al., 1976). Because of the diversity of sizes and the various treatments used to induce these fibres, it is unlikely that they are all composed of the same material. Their identity is unknown but their presence indicates the complexity and diversity of the nucleoplasm.

VIII.   MITOSIS ASSOCIATED NUCLEAR MOVEMENTS

Fungal nuclei may exhibit a number of types of movement.

Interphase nuclei may move through hyphae at about the same
rate as the hyphal tip advances (Girbardt, 1968, Snider, 1968,
Niederpruem, 1969). The rate of this type of movement in
Schizophyllum and Polystictus is in the order of 2 - 4 μm/min
at about 22° C (Snider, 1968, Girabardt, 1968). Nuclei may
also migrate through hyphae over long distances (60 μm to many
mm for Fomes and Schizophyllum respectively; Wilson & Aist,
1967, Snider, 1968) at rates from 8 to 300 μm/min (Snider,
1965). These examples, where the nuclei either approximately
hold their station with respect to the hyphal tip, or migrate
over substantial distances, have little direct connection with
mitosis although at least the former, and probably the latter,
cease during mitosis (e.g. Girbardt, 1968). However during
mitosis nuclei may oscillate with rates of movement in the
order of 30 μm/min (Girbardt, 1968), move over short distances
(up to 25 μm, Aist, 1969) and, at telophase and immediately
thereafter, the daughter nuclei may separate rapidly (Girbardt, 1968, Snider, 1968, Aist, 1969, Neiderpruem, 1969, Heath
& Heath, 1978) with rates for individual daughter nuclei in
the vicinity of 10 μm/min (Snider, 1968) to 16 μm/min (Girbardt, 1968). It is these relatively rapid movements occurring
during, or shortly after, mitosis which will be discussed
further here. Because they have comparable rates to the interphase nuclear migration they may well have a comparable
force mechanism. However, the tip related movements are not
only slower but also appear to have a different Q_{10} (1.4 vs.
6) (Snider, 1968) indicating that the mechanism for their
movement may well be different.

The force generating mechanism for the mitosis associated
movements most probably involves some sort of interaction between cytoplasmic microtubules, the nuclear envelope or NAO,
and the cytoplasm. The evidence for this statement is that at
least in Fusarium (Aist & Williams, 1972) and Uromyces (Heath
& Heath, 1976) (but not Puccinia [Harder, 1976a,b]) cytoplasmic microtubules radiating from the NAOs substantially increase in number at anaphase-telophase at which time the
nuclei are elongating prior to the separation of the daughter
nuclei. This increase suggests either a role in nuclear elongation or daughter nucleus movement, or both. In Polystictus,
Girbardt (1968) has also shown that abundant NAO-based cytoplasmic microtubules are present during mitosis associated
nuclear movements. That the cytoplasmic microtubules interact
with the cytoplasm may be indicated by Girbardt's (1968)

observation that the cytoplasm, and its contained organelles, move with the oscillating nuclei for up to 10 μm away from the spindle poles. The nature of this interaction is totally obscure. However changes in the cytoplasm around mitotic nuclei have been noted in <u>Basidiobolus</u> by Robinow (1963). Evidently the process of mitosis affects more than just the nucleus and the spindle but details of these cytoplasmic changes, their possible interaction with the NAO associated cytoplasmic microtubules and their involvement in mitosis associated movements, all remain topics for future research.

In the above situations the cytoplasmic microtubules radiate from the NAO in such a way that any force that they may transmit to the nuclei is presumably transmitted via these organelles. However in some oomycetes there appears to be another arrangement whereby the cytoplasmic microtubules interact laterally with the nuclear envelope (Heath & Greenwood, 1970, Heath, 1974b,c); thus force is presumably either generated by this interaction (sliding of membrane along the tubule) or transmitted from a microtubule-cytoplasmic interaction. Similar microtubule membrane interactions were discussed more in Section V A and in Heath (1975d). If this interaction were merely transmitting the force there would be greater homology with those fungi where the microtubules abut the NAO.

All of the above examples refer to hyphal fungi. However in the yeast forms there are also examples of mitosis associated nuclear movements whereby the nuclei variously migrate into and out of the bud prior to, or after, mitosis. Whilst rates for these movements have not been established, it is clear that in at least some genera there are NAO associated cytoplasmic microtubules which precede the migrating nuclei (McCully & Robinow, 1972b, Poon & Day, 1976a,b). Similar cytoplasmic microtubules also appear to be involved in nuclear fusion during mating of <u>Saccharomyces</u> (Byers & Goetsch, 1975). Thus it seems probable that cytoplasmic microtubules associated with the NAO, or the nuclear envelope, are involved in mitosis associated movements in a wide array of fungi. Since Wilson and Aist (1967) showed that interphase migrating nuclei are preceded by their NAOs it is possible that this type of movement is also mediated by NAO linked cytoplasmic microtubules. However, the molecular basis for all nuclear movements in fungi needs much more work before any unambiguous conclusions can be drawn.

IX. CONTROL OF MITOSIS

Ideally one would like to know what it is in a cell which determines the time of initiation of mitosis and, once initiated, what controls the transition from one phase to the next. Whilst there are no definite answers to these questions there are a number of observations which give tantalizing clues. In plasmodial and mycelial fungi it is very commonly observed that mitosis is synchronized throughout a plasmodium or hypha (e.g. Rosenberger & Kessel, 1967, Van der Valk & Wessels, 1973). However, as noted by Ross (1967), Robinow and Caten (1969) and others, the degree of synchrony is not perfect and in some species (e.g. Ross, 1967, Aist, 1969) there appears to be a wave-like pattern to the asynchrony which suggests the presence of a diffusing control agent rather than some endogenous nuclear clock. The concept of a cytoplasmic control on mitosis is further supported by the observations which show that mitosis is either closely co-ordinated with cell size (e.g. Robinow, 1963, Fantes, 1977, Fiddy & Trinci, 1976) or that the nucleus to cytoplasmic volume ratio is kept constant (e.g. Sudbery & Grant, 1976, Fiddy & Trinci, 1976). The Myxomycete, Physarum has proved to be the organism on which most mitotic control work has been performed and it is in this genus that the best evidence for a diffusable cytoplasmic control factor(s) has been obtained although Robinow (1963) presented evidence for such an agent in Basidiobolus sometime earlier. In Physarum, Rusch, et al. (1966) showed that after fusing plasmodia which were at different phases of the cell cycle, the time of the next mitosis was intermediate between the predicted time of the parental plasmodia and was closer to the parent which had the larger volume at fusion. A similar experiment by Ross (1967) provided similar results for Perichaena. Oppenheim and Katzer (1971) demonstrated that a plasmodium homogenate, especially from G_2, contained a heat-labile, non-dialysable, lyophilizable component whose addition to a non-dividing plasmodium would cause the time of mitosis to be advanced relative to the time predicted without treatment. However, the nature of this agent is still obscure. Lovely and Threlfell (1976) have shown a large build-up of cyclic adenosine monophosphate (cAMP) immediately prior to mitosis in G_2 and it is tempting to think that this compound induces mitosis, but the direct experiment of artificially inducing mitosis by topical application of cAMP is not yet

reported. Also Watson and Berry (1977) report contrary data
for Saccharomyces, thus there appears to be no universal
trends in cAMP levels in cells. Furthermore cAMP does not
appear to be the agent obtained by Oppenheim and Katzir (1971)
mentioned above so, whilst cAMP may be involved, its role is
unclear. On the other hand Bradbury, et al. (1972a,b) have
shown that H_1 histone is phosphorylated just prior to mitosis
by a specific H_1 histone phosphokinase which can induce early
mitosis when applied to another plasmodium thus indicating
that it is important in controlling mitosis. The interactions
between this enzyme and cAMP are not yet reported. The true
picture may be yet more complex because Sudbery and Grant
(1975) note that a simple accumulation of an inducer to a
critical level does not fit their data on mitosis control in
Physarum. Kuehn (1972) has demonstrated a further potential
level of control by showing that Physarum contains a protein
kinase whose function is unknown, but whose sensitivity to
cAMP inhibition varies during the cell cycle. Because this
protein is maximally inhibited by cAMP at the time when cAMP
is at its highest concentration (G_2) it is an unlikely candidate for a role in initiating mitosis. The simple concept of
cytoplasmic synthesis of a necessary inducer is further con-
fused by the work of Telatnyk and Guttes (1972) who showed
that U.V. irradiation of Physarum induces degeneration of some
damaged nuclei and simultaneous acceleration of mitosis of remaining "healthy" nuclei, a phenomenon which they interpret as
indicating scavenging of some nuclear components to accelerate
the process in the good nuclei.

   Whilst, as indicated above, the data on Physarum has not
given a very clear indication of how mitosis is initiated, it
is far ahead of any work on the other "fungi". Any student of
mycology is well aware of the beautiful way in which homobasidiomycetes co-ordinate nuclear division, nuclear movement and
septation yet there has been virtually no work on how this co-
ordination, especially the induction of mitosis, is achieved.
Similarly in the other "fungi" the only work of any substance
is that of Olson (1973) who studied accelerated initiation of
mitosis in Allomyces germlings. He observed that Na^+, K^+,
Ca^{++}, Mg^{++}, adenine, adenosine, guanine, guanosine, adenosine
tri-, di- and mono-phosphates, and guanosine mono- and triphosphate at suitable concentrations could all accelerate the
time of mitosis whereas glucose and mannitol at high concentrations could block mitosis. The problem with this, and the

cAMP, work is of course that all these substances affect a multitude of cellular processes so that unambiguous interpretation of the results becomes almost impossible (however, see Olson [1973] for a possible synthetic conclusion from his data).

As seen above the data on induction of mitosis is sparce and has not yet explained the phenomenon. Unfortunately, once induced, the means by which progress through mitosis is controlled is 'even more obscure. This reviewer knows of no helpful data on any fungus. Certainly the possibility that the nuclear envelope and the various intranuclear vesicles (mentioned in Sections III and Vc) function in spindle ion concentration control in a manner analogous to the muscle sarcoplasmic reticulum is an intriguing and tempting one but no useful data has been obtained to support this hypothesis. Similarly it has been noted that there may be vesicles specifically associated with the NAO during mitosis (Moens, 1976, Heath & Heath, 1976, Harder, 1976b). It is tempting to believe that they may have a regulatory role in mitosis but again there is no real evidence to support this speculation.

Clearly the whole question of control of mitosis in fungi is an important one to which any contribution is likely to be worthwhile in the existing desert of information.

X. SUMMARY

Nuclear division in the fungi presents a number of intriguing variations on the basic theme of mitosis. There is no compelling reason to believe that any of these variations indicate fundamentally different force producing mechanisms for either spindle elongation or chromosome to pole movements. However since the molecular basis for such mechanisms is essentially unknown in any mitotic system this point is not very helpful. A number of aspects of fungal mitosis suggest that further work will be particularly rewarding. For example small spindles are easier to analyze with the electron microscope than large ones. The remarkable synchrony of myxomycete plasmodial mitoses has already been useful for studies on control of mitosis. Intranuclear spindles can be analyzed by various means with less fear of cytoplasmic contamination than open spindles. There are undoubtedly other advantages which will become apparent. However, as with all investigations,

studies on fungal mitosis must be carried out with considerable regard to detail and with a firm grasp of the current
state of the art in other organisms; otherwise many important
points which could be easily resolved will go unnoticed. The
variations from "higher" mitotic systems, and between different fungi, are of course worth further analysis for the contributions which they can make to the understanding of mitosis
evolution and fungal phylogeny. However many more examples
need to be analyzed before sound conclusions can be drawn.

ACKNOWLEDGMENTS

A number of people have made important contributions to
this work. Drs. B.R. Oakley, A. Forer and P.B. Moens have
provided frequent stimulating discussions. Dr. M.C. Heath
provided a constant source of encouragement and critical discussion, as well as a review of the draught manuscript.
F. Holder typed the first draught and D. Gunning had the oner-
ous task of getting the manuscript to camera ready copy, a job
she performed with outstanding skill and patience. Finally,
the National Research Council of Canada provided the financial
assistance essential to the production of the unpublished data
referred to in this paper. All of these contributions were
most helpful and are very much appreciated. However, the
errors are of course all my own.

REFERENCES

Aist, J.R. (1969). J. Cell Biol. 40, 120-135.
Aist, J.R., and Williams, P.H. (1972). J. Cell Biol. 55,
    368-389.
Aldrich, H.C. (1967). Mycologia 59, 127-148.
Aldrich, H.C. (1969). Am. J. Bot. 56, 290-299.
Aldrich, H.C., and Carroll, G. (1971). Mycologia 63,
    308-316.
Allen, E.D., Lowry, R.J., and Sussman, A.S. (1974). J. Ult.
    Res. 48, 455-464.
Arrighi, F.E. (1974). In "The Cell Nucleus" (H. Busch, ed.),
    Vol. II, pp. 1-33. Academic Press, New York.
Bajer, A. (1973). Cytobios 8, 139-160.
Bajer, A., and Mole-Bajer, J. (1972). In "Spindle Dynamics

164   I. Brent Heath

and Chromosome Movements. International Review of Cytology Supplement 3" (G.H. Bourne, J.F. Danielli and K.W. Jeon, eds.), pp. 1-271. Academic Press, New York.

Bakerspigel, A. (1958). Am. J. Bot. 45, 404-410.

Bakerspigel, A. (1959). Am. J. Bot. 46, 180-190.

Banerjee, S., Kelleher, J.K., and Margulis, L. (1975). Cytobios 12, 171-178.

Beck, D.P., Decker, G.L., and Greenawalt, J.W. (1970). J. Ultrastruct. Res. 33, 245-251.

Beckett, A., and Crawford, R.M. (1970). J. Gen. Microbiol. 63, 269-280.

Bent, K.J., and Moore, R.H. (1966). In "Symposium of the Society for General Microbiology XVI. Biochemical Studies of Antimicrobial Drugs" (B.A. Newton and P.E. Reynolds, eds.), pp. 82-110. Cambridge University Press, London.

Berezney, R., and Coffey, D.S. (1977). J. Cell Biol. 73, 616-637.

Bhattacharyya, B., and Wolff, J. (1975). J. Biol. Chem. 250, 7639-7646.

Bibring, T., Baxandall, J., Denslow, S., and Walker, B. (1976). J. Cell Biol. 69, 301-312.

Bod, P., Szarka, E., Gyimesi, J., Horvath, G., Vajna-Méhesfalvi, Z., and Horvath, I. (1973). J. Antibiotics 26, 101-103.

Bollen, G.J., and Fuchs, A. (1970). Neth. J. Plant Path. 76, 299-312.

Boothroyd, B., Napier, E.J., and Somerfield, G.A. (1961). Biochem. J. 80, 34-37.

Borisy, G.G. (1977). Discussion at Mitosis Workshop, Heidelberg, April, 1977.

Borisy, G.G., Peterson, J.B., Hyams, J.S., and Ris, H. (1975). J. Cell Biol. 67, 38a.

Bourgois, J.-J., Bronchart, R., Deltour, R., and De Barsy, T. (1977). Pestic. Bichem. Physiol. 7, 97-106.

Bradbury, E.M., Inglis, R.J., and Matthews, H.R. (1974). Nature 247, 257-261.

Bradbury, E.M., Inglis, R.J., Mathews, H.R., and Langan, T.A. (1974). Nature 249, 553-555.

Braselton, J.P., Miller, C.E., and Pechak, D.G. (1975). Amer. J. Bot. 62, 349-358.

Brinkley, B.R., and Cartwright, J. (1970). J. Cell Biol. 47, 25a.

Brinkley, B.R., Stubblefield, E., and Hsu, T.C. (1967). J.

Ultrastruct. Res. 19, 1-18.

Burns, R.G. (1973). Exptl Cell Res. 81, 285-292.

Byers, B., and Goetsch, L. (1975a). Proc. Nat. Acad. Sci. U.S.A. 72, 5056-5060.

Byers, B., and Goetsch, L. (1975b). J. Bact. 124, 511-523.

Clarke, M., Schatten, G., Mazia, D., and Spudich, J.A. (1975). Proc. Nat. Acad. Sci. U.S.A. 72, 1758-1762.

Coffey, M.D., Palevitz, B.A., and Allen, P.J. (1972). Can. J. Bot. 50, 231-240.

Cortese, F., Bhattacharyya, B., and Wolff, J. (1977). J. Biol. Chem. 252, 1134-1140.

Coss, R.A., Bloodgood, R.A., Brower, D.L., Pickett-Heaps, J. D., and McIntosh, J.R. (1975). Exptl Cell Res. 92, 394-398.

Crackower, S.H.B. (1972). Can. J. Microbiol. 18, 683-687.

Craigie, J.H. (1959). Can. J. Bot. 37, 843-859.

Dancker, P., Low, I., Hasselbach, W., and Wieland, T. (1975). Biochim. Biophys. Acta. 400, 407-414.

Davidse, L.C. (1973). Pestic. Biochem. Physiol. 3, 317-325.

Davidse, L.C. (1975). In "Microtubules and Microtubule Inhibitors" (M. Borgers and M. De Brabander, eds.), pp.483-495. North-Holland, Amsterdam.

Davidse, L.C. (1976). Pestic. Biochem. Physiol. 6, 538-546.

Davidse, L.C., and Flach, W. (1977). J. Cell Biol. 72, 174-193.

Day, A.W. (1972). Can.J. Bot. 50, 1337-1347.

De Brabander, M., Van de Veire, R., Aerts, F., Geuens, G., Borgers, M., and Desplenter, L. (1975). In "Microtubules and Microtubule Inhibitors" (M. Borgers and M. De Brabander, eds.), pp. 509-521. North Holland, Amsterdam.

De Brabander, M., Van de Veire, R.M.L., Aerts, F.E.M., Borgers, M., and Janssen, P.A.J. (1976). Cancer Res. 36, 905-916.

Dietz, R. (1972). Chromosoma 38, 11-76.

Dippell, R.V. (1976). J. Cell Biol. 69, 622-637.

Dunkle, L.D., Wergin, W.P., and Allen, P.J. (1970). Can. J. Bot. 48, 1693-1695.

Dykstra, M.J. (1976). Protoplasma 87, 347-359.

Edgington, L.V., Khew, K.L., and Barron, G.L. (1971). Phytopathology 61, 42-44.

Eigsti, O.J., and Dustin, P. (1955). "Colchicine in Agriculture, Medicine, Biology and Chemistry". Iowa State College Press, Ames, Iowa.

Ellzey, J.T. (1974). Mycologia 66, 32-47.

Ellzey, J., Huizar, E., and Yanez, D. (1976). Arch. Microbiol. 107, 113-114.

Elsner, P.R., Vander Molen, G.E., Horton, J.C., and Bowen, C.C. (1970). Phytopathology 60, 1765-1772.

Fantes, P.A. (1977). J. Cell Sci. 24, 51-67.

Faulkner, B.M. (1967). Aspergillus News Letter 8, 18.

Felden, R.A., Sanders, M.M., and Morris, N.R. (1976). J. Cell Biol. 68, 430-439.

Feldherr, C.M. (1972). In "Advances in Cell and Molecular Biology" (E.J. Du Praw, ed.), Vol. II, pp. 273-309. Academic Press, New York.

Fiddy, C., and Trinci, A.P.J. (1976). J. Gen. Microbiol. 97, 169-184.

Fischer, P., Binder, M., and Wintersberger, U. (1975). Exptl Cell Res. 96, 15-22.

Forer, A. (1974). In "Cell Cycle Controls" (G.M. Padilla, I.L. Cameron, and A.M. Zimmerman, eds.), pp. 319-335. Academic Press, New York.

Franco, J., Johns, E.W., and Navlet, J.M. (1974). Eur. J. Biochem. 45, 83-89.

Franke, W.W., and Reau, P. (1973). Arch. Mikrobiol. 90, 121-129.

Fuge, H. (1974). Chromosoma 45, 245-260.

Fuller, M.S. (1976). Int. Rev. Cytol. 45, 113-153.

Fuller, M.S., and Reichle, R. (1965). Mycologia 57, 946-961.

Fulton, C. (1971). In "Origin and Continuity of Cell Organelles" (J. Reinert and H. Urspring, eds.), pp. 170-221. Springer-Verlag, New York.

Furtado, J.S., and Oliver, L.S. (1970). Cytobiologie 2, 200-219.

Garber, R.C., and Aist, J.R. (1977). In "Abstracts, 2nd International Mycological Congress" (H.E. Bigelow and E.G. Simmons, eds.), p.217.

Girbardt, M. (1961). Exptl Cell Res. 23, 181-194.

Girbardt, M. (1962). Planta 58, 1-21.

Girbardt, M. (1968). In "Aspects of Cell Motility" (P.L. Miller, ed.), Symp. Soc. Exp. Biol. 22, pp.249-259. Cambridge Univ. Press, London and New York.

Girbardt, M. (1971). J. Cell Sci. 2, 453-473.

Girbardt, M. (1977). Discussion at Mitosis Workshop, Heidelberg, April, 1977.

Girbardt, M., and Hädrich, H. (1975). Z. Allg. Mikrobiol.

15, 157-173.

Gleason, F.H. (1973). Cytobios 8, 185-187.

Goldstein, L. (1974). In "The Cell Nucleus" (H. Busch, ed.), Vol. I, pp. 388-440. Academic Press, New York.

Goode, D. (1975). BioSystems 7, 318-325.

Goodman, E.M., and Ritter, H. (1969). Arch. Protistenk. 111, 161-169.

Gordon, C.N. (1977). J. Cell Sci. 24, 81-94.

Gould, R.R., and Borisy, G.G. (1977). J. Cell Biol. 73, 601-615.

Gull, K., and Newsam, R.J. (1975). Protoplasma 83, 247-257.

Gull, K., and Newsam, R.J. (1976). Protoplasma 90, 343-352.

Gull, K., and Trinci, A.P.J. (1973). Nature 244, 292-294.

Gull, K., and Trinci, A.P.J. (1974). Trans. Brit. Mycol. Soc. 63, 457-460.

Gull, K., and Trinci, A.P.J. (1974a). Arch. Mikrobiol. 95, 57-65.

Gull, K., and Trinci, A.P.J. (1974b). Protoplasma 81, 37-48.

Guth, E., Hashimoto, T., and Conti, S.F. (1972). J. Bacteriol. 109, 869-880.

Guttes, S., Guttes, E., and Ellis, R.A. (1968). J. Ultrastruct. Res. 22, 508-529.

Haber, J.E., Peloquin, J.G., Halvorson, H.O., and Borisy, G. G. (1972). J. Cell Biol. 55, 355-367.

Hammerschlag, R.S., and Sisler, H.O. (1973). Pestic. Biochem. Physiol. 3, 42-54.

Harder, D.E. (1976a). Can. J. Bot. 54, 981-994.

Harder, D.E. (1976b). Can. J. Bot. 54, 995-1009.

Harris, P. (1975). Exptl Cell Res. 94, 409-425.

Hartman, H., Puma, J.P., and Gurney, T. (1974). J. Cell Sci. 16, 241-259.

Haskins, E.F. (1966). Chromosoma 56, 95-100.

Hastie, A.C. (1962). J. Gen. Microbiol. 27, 373-382.

Hauser, M., Beinbrech, G., Gröschel-Stewart, U., and Jockusch, B.M. (1975). Exptl Cell Res. 95, 127-135.

Heath, I.B. (1969). Ph.D. Thesis. Imperial College, University of London.

Heath, I.B. (1974a). In "The Cell Nucleus" (H. Busch, ed.), Vol. II, pp. 487-515. Academic Press, New York.

Heath, I.B. (1974b). J. Cell Biol. 60, 204-220.

Heath, I.B. (1974c). Mycologia 66, 354-359.

Heath, I.B. (1975a). BioSystems 7, 351-359.

Heath, I.B. (1975b). Protoplasma 85, 147-176.

Heath, I.B.   (1975c).   Protoplasma 85, 177-192.
Heath, I.B.   (1975d).   In "Proc. 1st Intersectional Congress of Int. Assoc. Microbiol. Socs." (T. Hasegawa, ed.), Vol. 2, pp. 92-106. Tokyo University Press, Tokyo.
Heath, I.B.   (1977).   In "Abstracts, 2nd International Mycological Congress" (H.E. Bigelow and E.G. Simmons, eds.), p. 275.
Heath, I.B., and Greenwood, A.D. (1968). J. gen. Microbiol. 53, 287-289.
Heath, I.B., and Greenwood, A.D. (1970). J. gen. Microbiol. 62, 139-148.
Heath, I.B., Greenwood, A.D., and Griffiths, H.B. (1970). J. Cell Sci. 7, 445-461.
Heath, I.B., and Heath, M.C. (1976). J. Cell Biol. 70, 592-607.
Heath, M.C., and Heath, I.B. (1978). Can. J. Bot. 56. In press.
Heideman, S.R., Sander, G., and Kirschner, M.W. (1977). Cell 10, 337-350.
Hemmes, D.E., and Hohl, H.R. (1973). Can. J. Bot. 51, 1673-1675.
Hepler, P.K., and Palevitz, B.A. (1974). Ann. Rev. Plant Physiol. 25, 309-362.
Hinchee, A.A. (1976). Ph.D. Thesis, University of Washington, Seattle.
Hoch, H.C., and Mitchell, J.E. (1972). Protoplasma 75, 113-138.
Hoebeke, J., Van Nijen, G., and De Brabander, M. (1976). Biochem. Biophys. Res. Comm. 69, 319-324.
Horgen, P.A., Nagao, R.T., Chia, L.S.Y., and Key, J.L. (1973). Arch. Mikrobiol. 94, 249-258.
Hotta, Y., and Shepard, J. (1973). Mol. Gen. Genet. 122, 243-260.
Howard, K.L., and Moore, R.T. (1970). Bot. Gaz. 131, 311-336.
Howard, R.J., and Aist, J.R. (1977). Protoplasma 92, 195-210.
Hsiang, M.W., and Cole, R.D. (1973). J. Biol. Chem. 248, 2007-2013.
Huang, H.C., Tinline, R.D., and Fowke, L.C. (1975). Can. J. Bot. 53, 403-414.
Ichida, A.A., and Fuller, M.S. (1968). Mycologia 60, 141-155.

Inoué, S., and Sato, H. (1967). J. Gen. Physiol. 50, 259-292.

Jensen, C., and Bajer, A. (1973). Chromosoma 44, 73-89.

Jockusch, B.M. (1973). Ber. Deutsch. Bot. Ges. 86, 39-54.

Jockusch, B.M., Becker, M., Hindennach, I., and Jockusch, H. (1974). Exptl Cell Res. 89, 241-246.

Jockusch, B.M., Ryser, U., and Behnke, O. (1973). Exptl Cell Res. 76, 464-466.

Jorgensen, A.O., and Heywood, S.M. (1974). Proc. Nat. Acad. Sci. U.S.A. 71, 4278-4282.

Käfer, E. (1961). Genetics 46, 1581-1609.

Kazama, F.Y. (1974). Protoplasma 82, 155-175.

Kerr, S.J. (1967). J. Protozool. 14, 439-445.

Keskin, B. (1971). Arch. Mikrobiol. 77, 344-348.

Khan, S.R. (1976). Can. J. Bot. 54, 168-172.

Kiermayer, O., and Fedtke, C. (1977). Protoplasma 92, 163-166.

Koevenig, J.L. (1964). Mycologia 56, 170-184.

Kohli, J., Hottinger, H., Munz, P., Strauss, A., and Thuriaux, P. (1977). Genetics 87, 471-489.

Kubai, D.F. (1973). J. Cell Sci. 13, 511-552.

Kubai, D.F. (1975). Int. Rev. Cytol. 43, 167-227.

Kuehn, G.D. (1972). Biochem. Biophys. Res. Comm. 49, 414-419.

Kumari, L., Decallonne, J.R., and Meyer, J.A. (1977). Pestic. Biochem. Physiol. 7, 273-282.

Kunkel, W., and Hädrich, H. (1977). Protoplasma 92, 311-323.

Kuzmich, M.J., and Zimmerman, A.M. (1972). Exptl Cell Res. 72, 441-452.

Lederberg, S., and Stetten, G. (1970). Science 168, 485-487.

Leighton, T.J., Dill, B.C., Stock, J.J., and Philips, C. (1971). Proc. Natl. Acad. Sci. U.S.A. 68, 677-680.

Lerbs, V. (1971). Arch. Mikrobiol. 77, 308-330.

Lerbs, V., and Thielke, C. (1969). Arch. Mikrobiol. 68, 95-98.

Lessie, P.E., and Lovett, J.S. (1968). Amer. J. Bot. 55, 220-236.

Lestourgeon, W.M., Forer, A., Yang, Y.-Z., Bertram, J.S., and Rusch, H.P. (1975). Biochim. Biophys. Acta. 379, 529-552.

Lovely, J.R., and Threlfell, R.J. (1976). Biochem. Biophys. Res. Comm. 71, 789-795.

Lu, B.C. (1967). J. Cell Sci. 2, 529-536.

Luftig, R.B., McMillan, P.N., Weatherbee, J.A. and Weihing, R.R. (1977). J. Histochem. Cytochem. 25, 175-187.

Maheshwari, R., Hildebrandt, A.C., and Allen, P.J. (1967). Can. J. Bot. 45, 447-450.

Mandelbaum-Shavit, F., Wolpert-De Filippes, M.K., and Johns, O.G. (1976). Biochem. Biophys. Res. Comm. 72, 47-54.

Margulis, L. (1973). Int. Rev. Cytol. 34, 333-361.

Matile, P.H., Moor, H., and Robinow, C.F. (1969). In "The Yeasts" (A.H. Rose and J.S. Harrison, eds.), Vol. 1, pp. 219-302. Academic Press, London and New York.

Mazia, D. (1961). In "The Cell" (J. Brachet and A.E. Mirskey, eds.), Vol. III, pp. 77-412. Academic Press, London and New York.

McCully, E.K., and Robinow, C.F. (1971). J. Cell Sci. 9, 475-507.

McCully, E.K., and Robinow, C.F. (1972a). J. Cell Sci. 11, 1-31.

McCully, E.K., and Robinow, C.F. (1972b). J. Cell Sci. 10, 857-881.

McCully, E.K., and Robinow, C.F. (1973). Arch. Mikrobiol. 94, 133-148.

McDonald, D.K. (1972). J. Phycol. 8, 156-165.

McIntosh, J.R., Cande, W.Z., and Snyder, J.A. (1975). In "Molecules and Cell Movement" (S. Inoué and R.E. Stephens, eds.), pp. 31-76. Raven Press, New York.

McIntosh, J.R., Hepler, P.K., and Van Wie, D.G. (1969). Nature 224, 659-663.

McKeen, W.E. (1972). Can. J. Microbiol. 18, 1915-1922.

McLaughlin, D.J. (1971). J. Cell Biol. 50, 737-745.

McManus, S.M.A., and Roth, L.E. (1968). Mycologia 60, 426-436.

McNitt, R. (1973). Can. J. Bot. 51, 2065-2074.

McNitt, R. (1974). Protoplasma 80, 91-108.

Mims, C.W. (1972). J. Gen. Microbiol. 71, 53-62.

Mims, C.W. (1977). Can. J. Bot. 55, 2319-2329.

Moens, P.B. (1973). Int. Rev. Cytol. 35, 117-134.

Moens, P.B. (1976). J. Cell Biol. 68, 113-122.

Moens, P.B., and Rapport, E. (1971). J. Cell Biol. 50, 344-361.

Mohberg, J. (1977). J. Cell Sci. 24, 95-108.

Moor, H. (1966). J. Cell Biol. 29, 153-156.

Moore, R.T.   (1964). Z. Zellforsch. 63, 921-937.

Moorman, G.W.   (1976). Mycologia 68, 902-909.

Mortimer, R.K., and Hawthorne, D.C.   (1969). In "The Yeasts"
     (A. Rose and J. Harrison, eds.), Vol. 1, pp. 386-460.
     Academic Press, London and New York.

Motta, J.J. (1969). Mycologia 61, 873-886.

Nicklas, R.B.   (1971). In "Advances in Cell Biology" (D.M.
     Prescott, L. Goldstein and E.H. McConkey, eds.), Vol. II,
     pp. 225-294. Appleton-Century-Crofts, New York.

Niederpruem, D.J.   (1969). Arch. Mikrobiol. 64, 387-395.

Oakley, B.R., and Dodge, J.D.   (1976). Protoplasma 88, 241-
     254.

Olive, L.S.   (1953). Botan. Rev. 19, 439-586.

Olmsted, J.B., and Borisy, G.G.   (1973). Ann. Rev. Biochem.
     42, 507-540.

Olson, L.W.   (1972). Arch. Mikrobiol. 84, 327-338.

Olson, L.W.   (1973). Arch. Mikrobiol. 91, 281-286.

Olson, L.W.   (1974a). C.r. Trav. Lab. Carlsberg 40, 113-124.

Olson, L.W.   (1974b). C.r. Trav. Lab. Carlsberg 40, 125-134.

Oppenheim, D.S., Hauschka, B.T., and McIntosh, J.R.   (1973).
     Exptl Cell Res. 79, 95-105.

Oppenheim, A., and Katzir, N.   (1971). Exptl Cell Res. 68,
     224-226.

Ormerod, W., Francis, S., and Margulis, L.   (1977). In
     "Abstracts, 2nd International Mycological Congress" (H.E.
     Bigelow and E.G. Simmons, eds.), p. 496.

Östergren, G.   (1950). Hereditas 36, 1-18.

Perkins, F.O.   (1970). J. Cell Sci. 6, 629-653.

Perkins, F.O., and Amon, J.P.   (1969). J. Protozool. 16,
     235-257.

Peterson, J.B., Gray, R.H., and Ris, H.   (1972). J. Cell
     Biol. 53, 837-841.

Peterson, J.B., and Ris, H.   (1976). J. Cell Sci. 22, 219-
     242.

Petzelt, C., and Auel, D.   (1977). Proc. Nat. Acad. Sci.
     U.S.A. 74, 1610-1613.

Pickett-Heaps, J.D.   (1969). Cytobios 1, 257-280.

Pickett-Heapt, J.D.   (1971). Cytobios 3, 205-214.

Pickett-Heaps, J.D.   (1972). Cytobios 5, 59-77.

Pickett-Heaps, J.D.   (1974). BioSystems 6, 37-48.

Poon, N.H., and Day, A.W.   (1974). Can.J. Microbiol. 20,
     739-746.

Poon, N.H., and Day, A.W.   (1976a). Can. J. Microbiol. 22,

507-522.

Poon, N.H., and Day, A.W. (1976b). Can. J. Microbiol. 22, 495-506.

Porter, D. (1972). Protoplasma 74, 427-448.

Powell, M.J. (1975). Can. J. Bot. 53, 627-646.

Raju, N.B., and Lu, B.C. (1973). J. Cell Sci. 12, 131-141.

Randell, J.T., and Disbrey, C. (1965). Proc. Roy. Soc. Ser. B. 162, 473-491.

Reichle, R.E., and Lichtwardt, R.W. (1972). Arch. Mikrobiol. 81, 103-125.

Richards, O.W. (1938). J. Bact. 36, 187-195.

Richmond, D.V., and Phillips, A. (1975). Pestic. Biochem. Physiol. 5, 367-379.

Rickards, G.K. (1975). Chromosoma 49, 407-455.

Robinow, C.F. (1957). Can. J. Microbiol. 3, 771-789.

Robinow, C.F. (1962). Arch. Mikrobiol. 42, 369-377.

Robinow, C.F. (1963). J. Cell Biol. 17, 123-152.

Robinow, C.F. (1977). Genetics 87, 491-497.

Robinow, C.F., and Bakerspigel, A. (1965). In "The Fungi, an Advanced Treatise" (G.C. Ainsworth and A.S. Sussman, eds.), Vol. I, pp. 119-142. Academic Press, New York.

Robinow, C.F., and Caten, C.E. (1969). J. Cell Sci. 5, 403-431.

Robinow, C.F., and Marak, J. (1966). J. Cell Biol. 29, 129-151.

Roobol, A., Gull, K., and Pogson, C.I. (1976). FEBS Letters 67, 248-251.

Roobol, A., Gull, K., and Pogson, C.I. (1977). FEBS Letters 75, 149-153.

Roos, U.-P. (1975a). J. Cell Biol. 64, 480-491.

Roos, U.-P. (1975b). J. Cell Sci. 18, 315-326.

Rosenberger, R.F., and Kessel, M. (1967). J. Bact. 94, 1464-1469.

Ross, I.K. (1967). Am. J. Bot. 54, 617-625.

Ross, I.K. (1968). Protoplasma 66, 173-184.

Rusch, H.P., Sachsenmaier, W., Behrens, K., and Gruter, V. (1966). J. Cell Biol. 31, 204-209.

Ryser, U. (1970). Z. Zellforsch. 110, 108-130.

Sakai, A., and Shigenaga, M. (1972). Chromosoma 37, 101-116.

Sansome, E., and Harris, B.J. (1962). Nature 196, 291-292.

Savile, D.B.O. (1939). Am. J. Bot. 26, 585-609.

Scheetz, R.W. (1972). Mycologia 64, 38-54.

Schrantz, J.P. (1967). C.R. Acad. Sci. Paris 264, 1274-1277.

Seiler, J.P. (1975). Mutation Res. 32, 151-168.

Setliff, E.C. (1977). In "Abstracts, 2nd International
     Mycological Congress" (H.E. Bigelow and E.G. Simmons,
     eds.), p. 607.

Setliff, E.C., Hoch, H.C., and Patton, R.F. (1974). Can. J.
     Bot. 52, 2323-2333.

Shapiro, H.S. (1976). In "Biological Handbooks. Cell Biol-
     ogy" (P.L. Altman and D.D. Katz, eds.), pp. 367-377.
     Federation of American Societies for Experimental Biol-
     ogy, Bethesda.

Sheir-Neiss, G., Nardi, R.V., Gealt, M.A., and Morris, N.R.
     (1976). Biochem. Biophys. Res. Comm. 69, 285-290.

Slifkin, M. (1967). Mycologia 59, 431-445.

Smith-Sonneborn, J., and Plaut, W. (1967). J. Cell Sci. 2,
     225-234.

Snyder, J.A., and McIntosh, J.R. (1976). Ann. Rev. Biochem.
     45, 699-720.

Stadler, J., and Franke, W.W. (1974). J. Cell Biol. 60,
     297-303.

Staron, T., and Allard, C. (1964). Phytriatrie-Phytopharm.
     13, 163-168.

Stearns, M.E., and Brown, D.L. (1976). J. Cell Biol. 70,
     242a.

Stiers, D.L. (1976). Can.J. Bot. 54, 1714-1723.

Styles, J.A., and Garner, R. (1974). Mutation Res. 26,
     177-187.

Subirana, J.A. (1968). J. Theor. Biol. 20, 117-123.

Sudbery, P.E., and Grant, W.D. (1975). Exptl Cell Res. 95,
     405-415.

Sun, N.C., and Bowen, C.C. (1972). Caryologia 25, 471-494.

Sundberg, W.J. (1977). In "Abstracts, 2nd International
     Mycological Congress" (H.E. Bigelow and E.G. Simmons,
     eds.), p. 644.

Tanaka, K. (1970). Protoplasma 70, 423-440.

Tanaka, K. (1973). J. Cell Biol. 57, 220-224.

Tanaka, K. (1977a). In "Abstracts, 2nd International
     Mycological Congress" (H.E. Bigelow and E.G. Simmons,
     eds.), p. 652.

Tanaka, K. (1977b). In "Growth and Differentiation in Micro-
     organisms" (T. Ishikawa, Y. Maruyama and H. Matsumiya,
     eds.), pp. 229-254. Tokyo University Press, Tokyo.

Thielke, C. (1973). Arch. Mikrobiol. 94, 341-350.

Thielke, C. (1974). Arch. Mikrobiol. 98, 225-237.

Tilney, L.G.   (1976).   J. Cell Biol. 69, 51-72.

Tilney, L.G., and Porter, K.R.   (1965).   J. Cell Biol. 34,
    327-343.

Tippit, D.H., and Pickett-Heaps, J.D.   (1977).   J. Cell Biol.
    73, 705-727.

Tonino, G., and Rozijn, T.   (1966).   Biochim. Biophys. Acta.
    124, 427-429.

Trinci, A.P.J., and Gull, K.   (1973).   Arch. Mikrobiol. 94,
    359-364.

Turian, G., and Oulevey, N.   (1971).   Cytobiologie 4, 250-261.

Upshall, A.   (1966).   Nature 209, 1113-1115.

Van Der Valk, P., and Wessels, J.G.H.   (1973).   Protoplasma
    78, 427-432.

Van Winkle, W.B., Biesele, J.T., and Wagner, R.P.   (1971).
    Can. J. Genet. Cytol. 13, 873-887.

Water, R.D., and Kleinsmith, L.J.   (1976).   Biochem. Biophys.
    Res. Comm. 70, 704-708.

Watson, D.C., and Berry, D.R.   (1977).   FEMS Microbiol.
    Letters 1, 175-178.

Weisenberg, R.   (1972).   Science 177, 1104-1105.

Wells, K.   (1970).   Mycologia 62, 761-790.

Whisler, H.C., and Travland, L.B.   (1973).   Arch. Protistenk.
    115, 69-74.

Wick, S.M., and Hepler, P.K.   (1976).   J. Cell Biol. 70, 209a.

Wilson, C.L., and Aist, J.R.   (1967).   Phytopathology 57,
    769-771.

Wilson, L., Anderson, K., Grisham, L., and Chin, D.   (1975).
    In "Microtubules and Microtubule Inhibitors" (M. Borgers
    and M. De Brabander, eds.), pp. 103-113.   North Holland,
    Amsterdam.

Winston, M., Johnson, E., Kelleher, J.K., Banerjee, S., and
    Margulis, L.   (1974).   Cytobios 9, 237-243.

Wintersberger, U., Binder, M., and Fischer, P.   (1975).
    Molec. gen. Genet. 142, 13-17.

Wintersberger, U., Smith, P., and Letnansky, K.   (1973).
    Eur. J. Biochem. 33, 123-130.

Witkins, E.R., and Berger, C.A.   (1944).   J. Hered. 35,
    130-133.

Witman, G.   (1973).   J. Cell Biol. 59, 366a.

Wolfe, J.   (1972).   In "Advances in Cell and Molecular
    Biology" (E.J. Du Praw, ed.), Vol. II, pp. 151-192.
    Academic Press, New York.

Wunderlich, F., and Herlan, G.   (1977).   J. Cell Biol. 73,

271-278.

Younger, K.B., Banerjee, S., Kelleher, J.K., Winston, M., and Margulis, L. (1972). J. Cell Sci. 11, 621-637.

Zackroff, R., Rosenfeld, A., and Weisenberg, R. (1975). J. Cell Biol. 67, 469a.

Zalokar, M. (1959). Am. J. Bot. 46, 602-610.

Zickler, D. (1970). Chromosoma 30, 287-304.

Zickler, D. (1971). C.R. Acad. Sci. Paris 273, 1687-1689.

Zickler, D. (1973). Histochemie 34, 227-238.

Zickler, D., and Olson, L.W. (1975). Chromosoma 50, 1-23.

Zimmerman, A.M. (1970). High Pressure Effects on Cellular Processes. Academic Press, New York.

ADDENDUM

   After this chapter was completed the author belatedly became aware of the paper by Bland and Lunney (1975) in which mitosis in the zygomycete, Pilobolus, is described. In this study many of the details are as described earlier for Mucor, Phycomyces and Conidiobolus. Thus the spindle poles have a small amount of intranuclear osmiophilic material adjacent to the nuclear envelope with just a little similar material on the outside also. Whether this external material qualifies as an NAO or not is an open question. As with the other Mucorales studied, the interaction between chromatin and spindle is ambiguous. Their Figure 9 (see also Fig. 7 of Kubai, this volume) has structures which look tantalizingly like chromosomal microtubules forming part of the short, developing spindle. Their preliminary figures of 14 microtubules in anaphase-telophase spindles and 16 - 36 in shorter spindles could be consistent with the existence of chromosomal microtubules but could equally be explained by two interdigitating groups which elongate and slide apart (with an undetected equatorial overlap zone) at anaphase-telophase. Clearly more detailed count data are required, together with microtubule tracking through serial sections, in order to resolve the question of the existence of chromosomal microtubules. Another point of interest from this paper is the suggestion that the remains of central part of the telophase spindle is excluded from the daughter nuclei. This is stated but not convincingly illustrated. If proven, it is the only record of lost nucleoplasm during a zygomycete mitosis. Finally, Pilobolus must be added to the list of organisms which have vesicles associated with the extranuclear polar regions during mitosis. Such vesicles are commented on, with a partial listing of occurrences, in Section IX. They also occur in the basidiomycetous yeasts examined by McCully and Robinow (1972a and b), in Polystictus (Girbardt, 1968), in Schizophyllum (Raudaskoski, 1970) and in Phycomyces (Franke & Reau, 1973). As noted earlier (Section IX) their role is unknown.

For comparison with Table 1, the data for Pilobolus mitosis can be summarized as follows, reading from left to right of the table: none, O, D, O, I, p, ?, +?, +?, +, b, ?.

References.

Bland, C.E., and Lunney, C.Z. (1975). Cytobiologie 11, 382-391.

Raudaskoski, M. (1970). Protoplasma 70, 415-422.

NUCLEAR DIVISION IN THE FUNGI

MITOSIS AND FUNGAL PHYLOGENY

Donna F. Kubai

Department of Zoology
Duke University
Durham, North Carolina
U.S.A.

I. INTRODUCTION

Phylogenetic taxonomy provides a shorthand notation for our deductions about the evolutionary relationships among organisms, and so must reflect profound understanding of various biological phenomena. In this paper, I consider the idea that comparative studies of mitosis can yield phylogenetic insights, not only in regard to relationships among the fungi, but also relative to more distant fungal antecedents.

Clearly, the task of reconstructing phyletic history is significantly easier when the evolutionary past is at least partially preserved, as it is in the fossil record for higher plants and animals. In considering the phylogeny of fungi, it has become something of a cliché to lament the limited fossil remains available for these organisms. Recently Tiffney and Barghoorn (1974) catalogued the fungal fossil record; and, although these authors maintain that the collection is not nearly so scant as has been thought (and will improve when more attention is devoted to paleomycology), they could derive only the most general phylogenetic clues from the presently available data.

The absence of a detailed fossil record is unquestionably a handicap in the study of phyletic history, but the bulk of evolution has affected structures and processes not susceptible to fossilization. Therefore, even for those organisms for which paleontological records are extensive, our "knowledge" of evolutionary history is heavily dependent on logical inference. Lacking a genuinely informative fossil record, mycologists are forced to base such extrapolation on their knowledge of extant organisms. Given that any living creature is the product of its evolutionary past, we are confident that the variations found among present-day organisms reflect their evolutionary origins. If, then, an array of variations is found to signal a degree of relatedness between living organisms and if these can be postulated to be the outcome of a reasonable evolutionary sequence, we should be able to reconstruct at least an outline of phylogenetic history. This rationale is the impetus for many current investigations; and comparisons of DNA base composition (for references, see Segal and Eylan, 1975), lipid profiles (Erwin, 1973; Southall, et al., 1977), metabolic pathways (LéJohn, 1974), cell wall chemistry (Bartnicki-Garcia, 1970), and cellular ultrastructure (Heath, 1974a, 1974c, 1975) promise to be helpful in assessing affinities among the fungi.

In any attempt to salvage evolutionary information from contemporary organisms, it is important to acknowledge two important facts of evolution: i) many of the evolutionary modifications leading to present-day structures or processes are irretrievably lost, having been supplanted by better-adapted variants, and ii) convergent evolution can lead to similar characters, thereby indicating closer relationships than in fact exist. With this in mind, we realize that no one set of comparisons can be taken as strong grounds for phylogenetic arguments. But, if parallel indications emerge from comparison of several characters, compelling arguments are possible.

Taking into account extinction of ill-adapted forms and convergent evolution, we might be entitled to little optimism that the comparative study of mitosis would prove helpful in understanding phylogeny. Since a mechanism for precise transmission of hereditary material is a sine qua non for any species, we could suppose that once the process of mitosis which is so exquisitely adaptive became established it supplanted all more primitive mechanisms of hereditary transmission. Thus, we would not be surprised to find that a highly evolved and quite invariant process of mitosis was the expedient for hereditary transmission in all existing organisms. Our knowledge of the disparate modes of hereditary transmission operative in the prokaryotes as contrasted with the eukaryotes (see Section II) assures us that this is not the case. In addition, numerous ultrastructural studies during the last decade show a rather surprising diversity in mitosis among the protists and fungi. These investigations have been reviewed by Fuller (1976), Heath (1974a), Kubai (1975), and Pickett-Heaps (1969, 1972, 1974). At a minimum, this heterogeneity can be taken as an indication of some evolutionary plasticity inherent in not-too-distant ancestral forms of nuclear division. Nonetheless, the certainty that

many of the most interesting and informative mitotic variants
have suffered extinction will complicate the task of understanding the significance of our findings.

There remains, in addition, the question of convergent
evolution. Savile (1968) has emphasized that convergence is
quite probable where organisms have only few ways of achiev-
ing essential ends. From this point of view, some degree of
convergent development of very similar mitotic mechanisms is
to be expected, and great caution must be exercised when
drawing phylogenetic conclusions from the study of mitosis.
Detailed evolutionary arguments will be possible only when
morphological, biochemical and ultrastructural evidence is in
agreement.

II. THE PROKARYOTE–EUKARYOTE TRANSITION

Recognition of the prokaryote-eukaryote dichotomy is a
useful frame of reference for our inquiry as to what sorts of
variants might prove phylogenetically significant in fungal
mitosis.

Prokaryotes are a class of organisms distinguished by a
mode of hereditary transmission which cannot be considered
mitotic (Chatton, 1937; Stanier, 1970; Stanier and van Niel,
1962). The genome of prokaryotes is relatively simple in
that all essential genetic information is contained in a
single DNA molecule. Consequently, these organisms have no
need for complicated machinery designed to coordinate the
distribution of several chromosomes to daughter cells. It
seems that by the mere attachment of replicated DNA molecules
to separate sites on the cell membrane, growth of the cell
surface between these sites could ensure passage of a complete genome to each daughter cell (Cuzin and Jacob, 1967;
Ryter, 1968; Liebowitz and Schaechter, 1975; see also the
discussion in Heath, 1974a).

In eukaryotes, total genetic information is contained in
several physically unlinked elements, the chromosomes, and
delivery of a complete set of these chromosomes to each
daughter cell must be accomplished with every cell division.
The precisely coordinated chromosome movement necessary for
this distribution is effected by mitosis, a process we recognize as distinctive of eukaryotes. In contrast to the situation in prokaryotes where genophore-membrane interaction
seems to be the intermediary for distribution of genetic
information, eukaryotes rely on chromosome-microtubule association as the basis for chromosome movement. During mitosis,
a specific site on each chromosome, the kinetochore, becomes
associated with microtubules and appropriate chromosome move-

ment takes place within a structure, the mitotic spindle,
which is in large part composed of microtubules. For discussions of our current understanding of mitosis, see Nicklas
(1971) and Forer (this volume).

Although opinions differ as to the precise way in which
prokaryotes and eukaryotes are related, it is generally held
that eukaryotes had prokaryotic progenitor(s) (Cohen, 1970,
1973; Flavell, 1972; Margulis, 1970; Raff and Mahler, 1972;
Raven, 1970; Sagen, 1967; Stanier, 1970; Uzzell and Spolsky,
1974). Since microtubules have not been found in prokaryotes
but are a ubiquitous component of eukaryotic cellular archi-
tecture (Porter, 1966), the acquisition of microtubules and
their gradual assumption of diverse functions is taken to be
a hallmark of early eukaryotic evolution (Stanier, 1970). In
the course of this evolution, then, we must expect that there
was a gradual transition from the membrane-mediated to
microtubule-mediated chromosome movement. We may therefore
wonder if any living organisms retain an element of membrane
involvement in chromosome movement. This idea has been discussed by Heath (1974a), Kubai (1975) and Pickett-Heaps (1969,
1972).

In typical eukaryotic mitosis, chromosome movement proceeds after the membranes of the nuclear envelope have dispersed. Although nuclear envelope-associated chromosome
movements have been recorded in prophase (Rickards, 1975)
(the initial mitotic stage during which the nuclear envelope
is yet intact), the significance of this activity relative to
appropriate chromosome distribution is not known. There are,
however, a large number of organisms in which the nuclear envelope remains either completely or nearly-completely intact
throughout mitosis (see Tables II and III in Kubai, 1975
and Table in Heath, this volume), and it is among these organisms where we might expect to find any evolutionary remnants of membrane-associated chromosome distribution. For
many of these, spindle formation and chromosome-spindle interactions are not recognizably unusual, albeit intranuclear,
and so we have no reason to think that the presence of the
nuclear envelope during mitosis ("closed mitosis") entails
exceptional chromosome behavior. On the other hand, for some
of the protists with closed mitosis, there is good evidence
that unconventional mitotic behavior derives from the presence of the nuclear envelope. It is organisms such as these
which may provide clues to the evolutionary development of
typical mitosis and thereby promote our understanding of the
phylogeny of protists.

Ultrastructural studies of nuclear division in two
groups of protists, the dinoflagellates and the hypermastigote flagellates, provide the most clear-cut evidence that

chromosome-nuclear envelope associations can be implicated in
the mechanism for chromosome distribution. In both of these
protozoan classes, the nuclear envelope persists during
mitosis and, most notable, the microtubules which become associated with the nucleus in preparation for division remain
outside of the nucleus.

In the dinoflagellates, extranuclear microtubules are
not even arrayed in a typical spindle-like configuration but
instead lie within one or several cytoplasmic channels pass-
ing through the dividing nucleus. For at least some dinoflagellates (Woloszynskia: Leadbeater and Dodge, 1967;
Crypthecodinium: Kubai and Ris, 1969; Blastodinium: Soyer,
1969, 1971), there is no evidence that microtubules interact
with the intranuclear chromosomes, but the demonstration that
there is a reasonably specific association of chromosomes and
the nuclear envelope (Fig. 1) has been interpreted as an indication that the mechanism of chromosome movement in these
species is similar to the membrane-mediated DNA distribution
of prokaryotes (Kubai and Ris, 1969; Soyer, 1969). In this
case, microtubules are seen as simply a device conferring
polarity on the spherical nucleus.

For other dinoflagellates (Amphidinium: Oakley and Dodge,
1974; Haplozoon: Siebert and West, 1974; Oodinium: Cachon and
Cachon, 1974, 1977; Syndinium: Ris and Kubai, 1974), extranuclear microtubules seem to be somewhat more directly involved in chromosome movement. In Syndinium, for example,
two distinct categories of microtubules are recognized: one
bundle running from pole-to-pole within a channel and others
which radiate from the poles. These latter microtubules abut
directly upon differentiated chromosome regions, the kinetochores, which protrude through pore-like openings in the
nuclear envelope (Fig. 2). In so far as a direct microtubule-
chromosome connection is involved in chromosome movement,
nuclear division in this group of dinoflagellates is more
like conventional mitosis than seems to be the case for the
species mentioned above. But, in Syndinium, the observation
that chromosomal microtubules do not shorten in conjunction
with poleward progress of the chromosomes implies that actual
chromosome movement is generated by elongation of the pole-to-
pole microtubules; chromosomes, then, are passively dragged
along with the separating poles to which they are anchored by
microtubules. This is in strong contrast to the usual mitotic course of events where chromosome movement is reflected
in actual shortening of chromosomal spindle fibers (which are
in part composed of microtubules). [A discussion of the
question of whether spindle fiber shortening actually gener-
ates the force for chromosome movement or is simply a necessary concommitant of chromosome movement is outside the scope

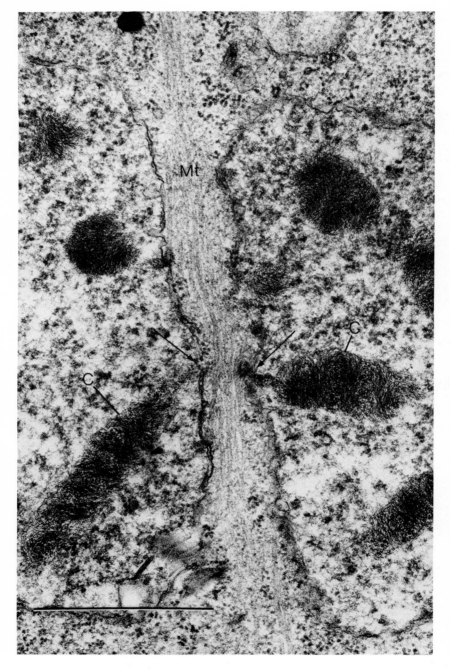

Fig. 1. See legend on next page.

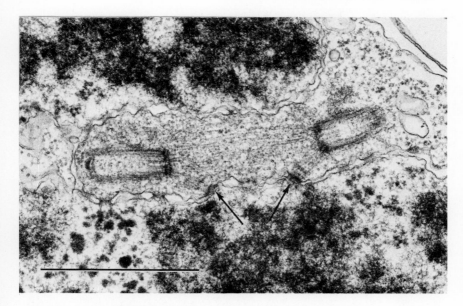

Fig. 2. <u>Syndinium</u> sp. Kinetochores (arrows) are in-
         serted in pore-like openings of the nuclear
         envelope and are connected to centrioles via
         cytoplasmic chromosomal microtubules.
         X 42,000. (From Ris and Kubai, 1974; repro-
         duced by permission of The Rockefeller
         University Press.)

of this presentation; the reader may refer to Forer (1974 and
this volume), Inoué and Ritter (1975) and McIntosh, <u>et al</u>.
(1975).]

Regarding the role of microtubules in dinoflagellate
nuclear division, the most important question remains open:
is there actually more than one type of division among these
protozoa--divisions in which microtubules never attach directly to chromosomes and those involving a more-or-less
direct microtubule connection to chromosomes? It will require further investigation to decide if the negative evi-

---

Fig. 1. <u>Crypthecodinium cohnii</u>. Microtubules (Mt) are
confined to cytoplasmic channels running through the nucleus.
Chromosomes (C) attach to the intact nuclear envelope
(arrows). X 47,500. (From Kubai and Ris, 1969; reproduced
by permission of The Rockefeller University Press.)

dence suggesting membrane-mediated chromosome movement in
certain species is valid (for discussion, see Kubai, 1975).
Nevertheless, present evidence suggests that in some dinoflagellates the chromosomes must either form or maintain a
rather specific association with the membranes of the nuclear
envelope in order to achieve appropriate chromosome distribution during mitosis.

In hypermastigotes, the indications of membrane-
associated chromosome movement are far less problematical.
Here, during division, a distinctly spindle-like extranuclear
microtubule array forms in preparation for division and eventually openings in the nuclear envelope allow direct interaction of kinetochores and microtubules (Fig. 3). These

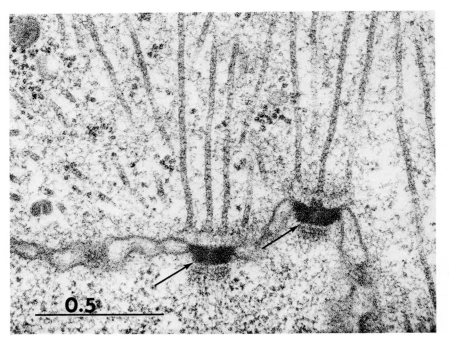

Fig. 3. _Trichonympha_ _agilis_. In late division
stages, kinetochores (arrows) are inserted
in pore-like openings of the nuclear envelope. Cytoplasmic chromosomal microtubules terminate at these kinetochores.
X 68,500. (From Kubai, 1973; reproduced
by permission of The Company of Biolo-
gists, Ltd.)

facts are the grounds for earlier workers' contentions that,

but for the extranuclear spindle placement, hypermastigote
mitosis conforms in all important particulars to conventional
mitosis (Cleveland, et al., 1934; Hollande and Carruette-
Valentin, 1970, 1971; Hollande and Valentin, 1968a, 1968b).
With a more detailed examination of spindle formation and
chromosome behavior in Trichonympha, however, it became evident that an important phase of chromosome movement occurs
well before kinetochores interact with microtubules (Kubai,
1973). Early in division, differentiated kinetochores are
intimately associated with the inner aspect of the nuclear
envelope (Fig. 4). These kinetochores occur in pairs (sister

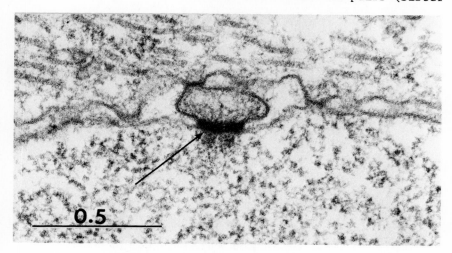

Fig. 4. Trichonympha agilis. In early division
        stages, kinetochores (arrow) are en-
        closed in outpocketings of the intact
        nuclear envelope and are thus incapable
        of direct interaction with cytoplasmic
        spindle microtubules. X 68,500. (From
        Kubai, 1973; reproduced by permission
        of The Company of Biologists, Ltd.)

chromosomes) distributed over the entire spherical nuclear
surface (Fig. 5). While the nuclear envelope remains unquestionably intact and thus forms a barrier to any direct
kinetochore-microtubule interaction, the paired kinetochores
disjoin and accumulate on a restricted portion of the nuclear
surface. These observations are incontrovertible evidence
that at least one phase of chromosome movement in hypermastigotes does not require direct microtubule-chromosome inter-
play and suggest that the membrane or the immediately under-

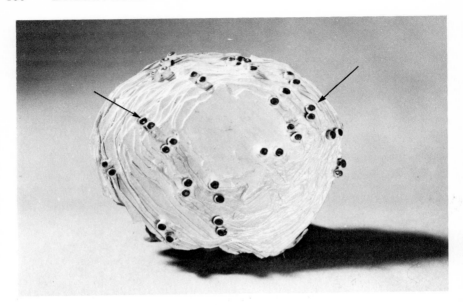

Fig. 5. <u>Trichonympha</u> <u>agilis</u>. While kinetochores
        are confined within the intact nuclear
        envelope as illustrated in Fig. 4, they
        occur in pairs (sister kinetochores)
        (arrows) which are scattered over the
        entire nuclear surface. (Reconstruction
        from serial sections.) (From Kubai,
        1973; reproduced by permission of
        The Company of Biologists, Ltd.)

lying cortex somehow mediates the movements that lead to dis-
junction of kinetochores and their redistribution on the nuclear surface. [For a study of spindle fiber involvement during later stages in a related hypermastigote, see Inoué and
Ritter (1975).] Although it is not obvious how such chromosome behavior promotes appropriate orientation of sister
kinetochores to opposite spindle poles, as would be required
if this division must deliver equivalent chromosome comple-
ments to each daughter cell, <u>Trichonympha</u> does show that membrane-associated chromosome movements can be an integral component of certain eukaryotic nuclear divisions.

III. ASPECTS OF FUNGAL MITOSIS

     A number of rather recent reviews have surveyed mitosis
in the fungi (Fuller, 1976; Heath, 1974a and this volume;

Kubai, 1975). Therefore these works may be consulted for the
specific ultrastructural information available on nuclear
envelope persistence, organelles occurring at the spindle
poles, kinetochore morphology and chromosome behavior for the
various species which have been studied. I will devote this
section to summary and evaluation of the peculiarities of
fungal mitosis which seem to me to be most likely to shed
some light upon the evolution of fungal mitosis and questions
of fungal phylogeny.

A. General Features

With few exceptions, the nuclear envelope remains essentially intact during mitosis in fungi; some gaps may appear,
e.g., in the region of spindle poles. Thus an intranuclear
spindle characterizes mitosis in the fungi.

Another general feature is the presence of differentiated structures on or associated with the nuclear envelope
during interphase. Since these organelles are found at
spindle poles during nuclear division, they are referred to
as spindle pole bodies [other designations have been used;
for a listing see Fuller (1976). See also Girbardt (this
volume) for discussion of this terminology.]. Because spindle pole bodies are the foci at which spindle microtubules
terminate, Pickett-Heaps (1969) suggested that they are probably microtubule organizing centers (i.e., structures capa-
ble of nucleating the polymerization of tubulin to form microtubules). In fact, the plaque-like spindle pole bodies in
Saccharomyces have been shown to nucleate microtubule growth
in vitro (Borisy, et al., 1975). In some Mastigomycotina and
some Zygomycotina, the spindle pole body appears as simply increased electron density of membrane(s) of the nuclear envelope and/or a layer of electron dense material appressed to
the inner nuclear membrane. In other Mastigomycotina, the
spindle pole body is identified with a centriole-associated
mass of amorphous material. In the remainder of the Eumycota,
more complicated structures appear; and, while these do assume various forms, there seems to be at least some degree of
consistency of structure in each subdivision. For example,
discoidal or plaque-like spindle pole bodies are common in
the Ascomycotina while globular spindle pole bodies seem to
be more characteristic of the Basidiomycotina. For a complete discussion of the variations in spindle pole body
structure and behavior, see Heath (this volume).

Aside from these two features, it is not possible to generalize about mitosis in the fungi. In a number of fungi,
present knowledge suggests that the intranuclear mitosis is

reasonably typical with condensed chromosomes aligning at the
spindle equator in a distinct metaphase plate and poleward
anaphase chromosome movement leading to distribution of sister chromosomes to daughter cells. Such apparently typical
mitotic figures are found among the Mastigomycotina, in Zygomycotina of the order Entomophthorales and in Ascomycotina of
the classes Pyrenomycetes and Discomycetes (see Tables I-III
in Kubai, 1975).

B. Anaphase Chromosome Behavior

     The impression that at least in some fungi mitosis does
not conform to our expectations for conventional mitosis
originated with light microscopic observations of the so-
called two-track or double-bar chromatin distribution in dividing cells. Such mitotic figures are observed frequently
in both the Ascomycotina and Basidiomycotina. In fact, Day
(1972) has pointed out that 70-80% of basidiomycetes for
which illustrations of nuclear division are available display
the two-track mitosis. As is so clear in the light micrographs published by Robinow and Caten (1969), these mitotic
figures involve bar-like masses of chromatin which align on
either side of a narrow central spindle. A transverse split
appears in each mass and movement of chromatin toward opposite spindle poles follows. The discrepencies between two-
track division figures and classical mitosis where individ-
ually recognizable chromosomes align across the spindle equator to form a metaphase plate is striking; and it is understandable that some investigators suggested that fungal
mitosis employs mechanisms grossly different than those operating in conventional mitosis. The discussions of Girbardt
(1968, 1971), Day (1972), Robinow and collaborators (McCully
and Robinow, 1973; Robinow and Caten, 1969), and Franke and
Reau (1973) are examples of the sorts of proposals which were
made. These focus on the fact that differentiated kinetochores connected to microtubular spindle fibers were only
seldom observed in the fungi (for a tabulation of recognizable
kinetochores in various fungi, see Kubai, 1975, Tables IV-VI).
     The idea that fungal mitosis might be pronouncedly dif-
ferent from classical mitosis is contradicted in large part
by the works of Aist and Williams (1972) and Setliff, et al.
(1974). Both of these investigations concern fungi which display the two-track chromatin configuration during mitosis:
Fusarium, an imperfect fungus with ascomycete affinities, and
Poria, a basidiomycete. As so clearly demonstrated by these
authors, the two-track image results because mitotic chromosomes become arranged around the periphery of the central

spindle, lying at various positions staggered along the spindle length (Fig. 6). Anaphase chromosome separation begins

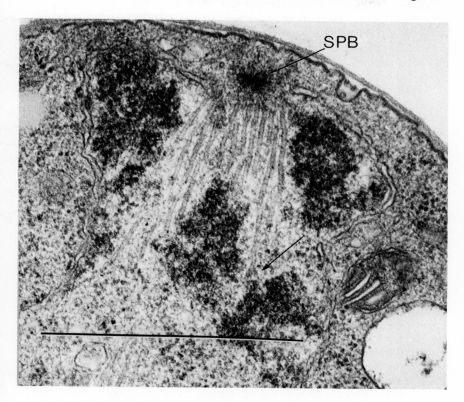

SPB

Fig. 6. <u>Poria latemarginata</u>. Anaphase-telophase
shows staggered chromosome placement.
Kinetochores (arrow) are minimally
differentiated, but the flared microtubule ending in these regions is distinctive. Spindle microtubules terminate near spindle pole bodies (SPB)
which lie at   polar openings of the
nuclear envelope. X 69,000. (From
Setliff, <u>et al</u>., 1974; reproduced by
permission of the National Research
Council of Canada from the Canadian
Journal of Botany, Volume 52, pp.
2323-2333, 1974.)

when chromosomes are in such a scattered array; and, thus,
chromosomes never become aligned in a well-defined plate at

the spindle equator. In light microscopy, then, optical sectioning of the rather cylindrical chromatin distribution produces the two-track images so often recorded. Moreover, differentiated kinetochores connecting to microtubules are demonstrable in these species. The work of Aist and Williams
(1972) is especially revealing in this context. These authors point out that choice of appropriate fixation conditions is particularly critical for the preservation of overall chromosome structure, and they found that small, minimally differentiated kinetochores and their associated microtubules are identifiable with confidence only through examination of serial sections (for an example of similar kinetochores, see Fig. 6). With this painstaking effort, they
showed that kinetochores occur in pairs, each kinetochore of
a pair being connected to the opposite spindle pole via a
single microtubule. As division progresses, the sister kinetochores disjoin, kinetochore microtubules shorten and chromosomes move toward the poles. Such poleward movement is indeed comparable to anaphase in classical mitosis. However,
the atypical pre-anaphase chromosome events which result in
staggered placement of chromosomes along the length of the
spindle (two-track division figures) is manifestly different
from the usual course of events where prometaphase chromosome
movements eventually bring all of the semi-independently
moving chromosomes to an equilibrium position at the spindle
equator. It seems possible that this difference signals a
critical modification in the mechanism of pre-anaphase chromosome movement for at least some of the fungi. We will return
to a discussion of this point (Section III.C.2.).

Thus, we find that anaphase chromosome movement in the
fungi may proceed after chromosomes have congressed at the
spindle equator, thereby displaying the classical metaphase
configuration, or they may be staggered along the length of
the spindle in the more dispersed array accounting for two-
track division figures. Heath's (1974b) study of Thraustotheca shows that in some fungi a condition intermediate between these may also occur: in this species, kinetochores are
distributed over a rather broad zone spanning the spindle
equator.

The studies of Aist and Williams (1972) and of Setliff,
et al. (1974) provide evidence for microtubule-mediated chromosome movement in a large group of fungi, but there remain
examples, especially among the zygomycetes and hemiascomycetes, where conventional electron microscopy yields essentially no information regarding the interactions between chromosomes and spindle. In the zygomycetes Mucor (McCully and
Robinow, 1973), Phycomyces (Franke and Reau, 1973) and Pilobolus (Bland and Lunney, 1975) and in the hemiascomycete

Saccharomyces (Byers and Goetsch, 1973) routinely fixed, thin-sectioned nuclei clearly display a central compact bundle of microtubules running from pole-to-pole (Figs. 7 and 8). How-

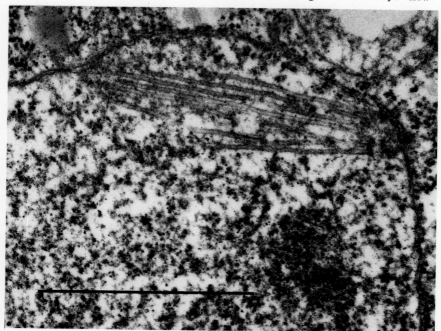

Fig. 7. Pilobolus crystallinus. The mitotic spindle consists of a sheaf of micro-tubules terminating near the nuclear envelope. Note absence of distin-guishable chromosomes and eccentric spindle position. X 56,900. (From Bland and Lunney, 1975; reproduced by permission of Wissenschaftliche Verlagsgesellschaft MBH.)

ever, no unequivocally identifiable chromosomal microtubules have been identified, nor indeed have individual chromosomes been recognized. In the face of such negative evidence, it was suggested that attachment of chromatin to the spindle pole bodies (McCully and Robinow, 1973) or to the nuclear en-velope (Franke and Reau, 1973) might be the agency for chro-mosome distribution. Elongation of the central spindle caus-ing separation of spindle pole bodies or of the chromatin-bearing sites on the nuclear envelope would then be responsi-ble for chromosome movement into daughter nuclei. Postulates

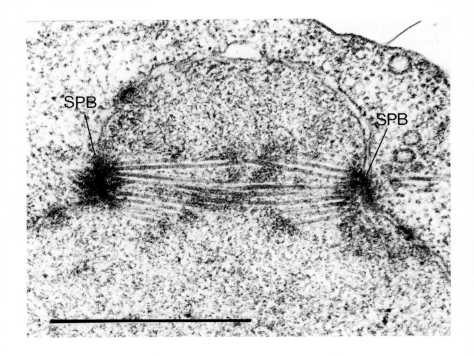

Fig. 8. Saccharomyces cerevisiae. When fully
formed, the mitotic spindle has, at
first, an eccentric position within
the nucleus. Later it will move to
a more central position. Microtubules span the distance between
separated spindle pole bodies (SPB)
and chromatin-like material lies
near the equator. X 53,500. (Courtesy of Joan B. Peterson.)

of this sort, which assume the utter absence of chromosomal
microtubules in the living cell, are difficult to accept in
light of our experience with fixation of microtubules. We
know that introduction of a special fixative was required before the ubiquitous occurrence of microtubules in eukaryotes
became apparent (Sabatini, et al., 1963; Porter, 1966).
Also, certain classes of microtubules are more susceptible
than others to breakdown by chemical treatment (Bajer and
Molè-Bajer, 1972; Brinkley, et al., 1967). It must also be
recalled that the minimally differentiated kinetochores and
the spindle microtubules with which they associate in
Fusarium, for example, require serial sectioning for their

demonstration, a technique that has been too seldom employed.

For at least one of the organisms in question, Saccharomyces, evidence for the existence of a distinct class of chromosomal microtubules is available. Peterson and Ris (1976) resorted to as yet little-exploited techniques--high voltage electron microscopy of serial thick sections and spreading of the nuclear contents on an air-water interface--in order to visualize the entire complement of microtubules in the dividing nucleus. By these means, they distinguished two sets of microtubules. One of them forms a bundle running from pole-to-pole and constitutes the central spindle. The other class includes microtubules which run from each pole only part way into the nuclear interior. These microtubules, thus, terminate far from the opposite poles; and, because their non-polar end is often seen to interact with chromatin-like fibers (see Fig. 8), they are taken to be chromosomal microtubules. The number of chromosomal microtubules counted in both haploid and diploid cells is consistent with the chromosome number determined in linkage studies and so these authors conclude that yeast chromosomes, in the form of relatively uncondensed chromatin fibers, are each associated with approximately one microtubule. Furthermore, their preparations show a reasonably distinct equatorial chromatin mass (Fig. 8) which splits into two masses that approach opposite poles as the chromosomal microtubules shorten. Peterson and Ris conclude that mitosis in this hemiascomycete is unexceptional except for the failure of chromosomes to condense into recognizable individuals.

The conclusion that mitosis in this fungus is unexceptional has been challenged by Gordon (1977) who also studied unconventionally prepared Saccharomyces. After ribonuclease digestion, nuclei are considerably less homogeneous than usual, and Gordon maintains that the inhomogeneity reveals chromatin masses which have been freed from obscuring RNA. Since the so-called chromatin bodies do not move progressively toward spindle poles as the nucleus elongates, Gordon takes issue with Peterson and Ris regarding chromosome movement. After comparing the electron micrographs obtained after such disparate treatments, it seems clear to me that there is some discrepency in the actual volume of material referred to as chromatin in these two publications. If all that Gordon calls chromatin is in fact chromosomal material, perhaps Peterson and Ris are visualizing only a portion of the total chromatin, specifically that which is most closely associated with chromosomal microtubules. And, if this is true, their idea that microtubule shortening acompanies the redistribution of chromatin is still valid. In any case, Gordon's techniques do not preserve any spindle structure and thus contribute little

to our understanding of the mechanisms of mitosis in these
fungi.

The fact that chromosomal microtubules and their interaction with uncondensed chromatin fibers were discovered in
Saccharomyces only after application of special techniques
makes it seem likely that similar indications will be found
in other Hemiascomycetes and the Zygomycotina if they are
more carefully examined (e.g., see the micrograph of the zygomycete Conidiobolus included in Heath's chapter in this volume).

To summarize, the most meticulous studies available to
us have shown that chromosomal microtubules do play a role in
anaphase chromosome movement in the majority of fungi; they
have been demonstrated in representatives of all the major
subdivisions of the Eumycota. There remain, therefore, no
compelling reasons to suppose that the mechanism of anaphase
chromosome distribution is extraordinary to any marked degree.

C. Preanaphase Behavior

It remains for us to consider preanaphase chromosome behavior in order to complete a comparison between fungal mitosis and classical mitosis. Pitifully few studies of fungal
mitosis devote even meager attention to study of prophase and
prometaphase. If we are to understand what might be the significance of the staggered array of chromosomes along the
length of the spindle in many fungi, the events leading up to
anaphase will have to be examined in detail. It is still
necessary to answer crucial questions regarding the manner in
which chromosomes become engaged with the spindle and how they
achieve their appropriate orientation with sister kinetochores directed toward opposite spindle poles. Fragments of
information which are available do suggest that features of
early spindle formation, chromosome-spindle interaction and
other aspects of chromosome behavior are unusual in some fungi.

1. Spindle Formation

The best available studies of spindle formation concern
the Oomycetes Thraustotheca (Heath, 1974b) and Saprolegnia
(Heath and Greenwood, 1968, 1970), the Chytridiomycetes
Entophlyctis (Powell, 1975), Phlyctochytrium (McNitt, 1973),
and Harpochytrium (Whisler and Travland, 1973) and the Hemiascomycetes Saccharomyces (Moens and Rapport, 1971; Byers and
Goetsch, 1973) and Schizosaccharomyces (McCully and Robinow,
1971). In each of these groups, a different mode of spindle

formation is illustrated.

In the Oomycetes, a slightly thickened nuclear envelope and an underlying lamina of electron dense material constitutes the spindle pole bodies. As mitosis begins, microtubules appear to connect the two spindle pole bodies, and their elongation probably contributes to the separation of spindle pole bodies on the nuclear surface. A separate group of microtubules is involved in formation of the intranuclear spindle.

In the Chytridiomycetes mentioned above, the microtubule organizing center which functions as the spindle pole body is extranuclear and associated with centrioles. These spindle pole bodies and the associated centrioles separate to opposite poles of the nucleus. At that time, gaps appear in the polar regions of the nuclear envelope and microtubules which have appeared around the spindle pole bodies are able to enter the nuclear space and engage in spindle formation.

In the Hemiascomycetes, spindle pole body structure is complex, e.g., in _Saccharomyces_ a highly differentiated plaque-like spindle pole body is inserted in a pore-like gap in the nuclear envelope. In these fungi, it is clear that the early separation of spindle pole bodies is not the result of microtubule elongation, since the fan-like radiation of microtubules which is associated with each spindle pole body terminates in the nuclear interior (Fig. 9). Thus, it is suggested that spindle pole body movement is in part a function of the expansion of the nuclear surface between two spindle pole bodies (McCully and Robinow, 1971). Later, the two arrays of microtubules interact and form a small eccentric spindle (Fig. 8). Such eccentric spindles are also found in mucoralean zygomycetes (Fig.7) and so it seems that a similar sequence of spindle formation occurs in this group. From this point on, elongation of the pole-to-pole microtubules would be a factor in spindle pole body separation and spindle elongation, events which eventually bring the spindle to a more central position in the nucleus.

These three examples demonstrate a striking range in spindle formation; they may be of significance not only as evidence that evolutionary modifications in spindle formation are found among the fungi but also because chromosome-spindle interaction might be affected by differing modes of spindle formation.

2. Relationship of Chromosomes and Spindle Pole Bodies

Numerous observations have suggested that a significant mass of fungal chromatin has a special relationship with the spindle pole body. In the interphase nucleus, connection be-

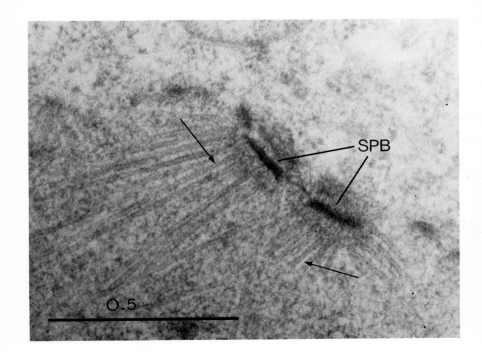

Fig. 9. Saccharomyces cerevisiae. As mi-
         tosis is initiated, fan-like arrays
         of microtubules (arrows) radiate
         from each of two adjacent spindle
         pole bodies (SPB). X 100,000. (From
         Byers and Goetsch, 1975; reproduced
         by permission of the authors.)

tween heterochromatin and a restricted region of the nuclear
envelope was noted even in early light microscopic investigations (e.g., Harper, 1905). In addition, we have the more
recent ultrastructural demonstrations that a mass of condensed chromatin-like material is associated with the inner
aspect of the nuclear envelope in the region of the spindle
pole body (e.g., Aist and Williams, 1972; Gull and Newsam,
1975; Harder, 1976a; Poon and Day, 1976a, 1976b) (Fig. 10).
     The possibility that a direct and intimate connection
between chromatin and the spindle pole body persists through
mitosis and bears the major responsibility for equitable
chromatin distribution to daughter nuclei has been enter-
tained (e.g., Girbardt, 1968, 1971). Such ideas became un-
tenable when the evidence discussed above showed that anaphase chromosome distribution in the majority of fungi is

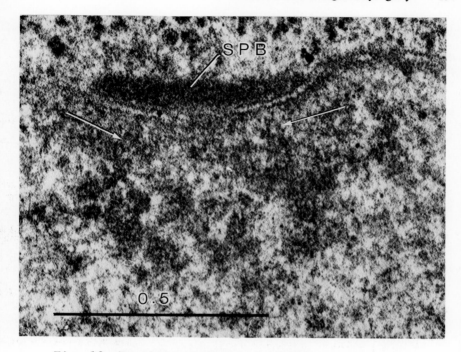

Fig. 10. Puccinia coronata. In interphase a
        mass of chromatin-like material
        (arrows) is closely associated with
        the nuclear envelope in the region
        which bears the extranuclear spindle
        pole body (SPB). X 114,000. (From
        Harder, 1976a; reproduced by per-
        mission of the National Research
        Council of Canada from the Canadian
        Journal of Botany, Volume 54, pp.
        981-994, 1976.)

dependent on a connection between chromosomes and chromosomal
microtubules. However, if, as can be supposed, these chromosomal microtubules directly link chromosomes and spindle pole
bodies, a slightly modified version of spindle pole body-
mediated chromosome movement can be envisioned. Aist and
Williams (1972) alluded to such a possibility when they suggested that the chromatin associated with spindle pole bodies
may be kinetochoric heterochromatin. If, then, kinetochore-
spindle pole body connection via microtubules is effected in
early prophase when spindle pole bodies are just beginning to
separate on the nuclear surface, movement of chromosomes directly dependent on spindle pole body movement might occur at

least in preanaphase stages.

In order to evaluate this idea and its possible consequences, it is important to know precisely when and how chromosomes become associated with the spindle. For the Eumycota,
the only published information on this important question is
provided by Heath and Greenwood (1970) who observed the earliest stages of spindle formation in Saprolegnia and by Heath
(1974b) who examined Thraustotheca. When spindle pole
bodies have just separated on the nuclear surface, chromosomal
microtubules already join chromosomes to the spindle pole
bodies (Fig. 11). As I have discussed elsewhere (Kubai,

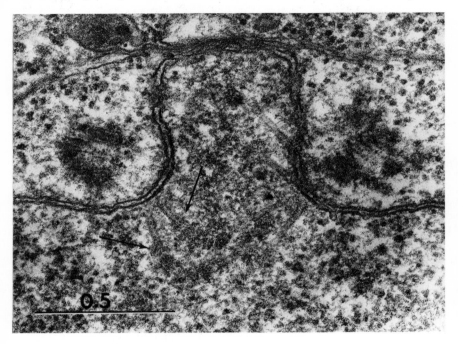

Fig. 11. Saprolegnia ferax. In an early divi-
          sion stage, when centrioles are not
          yet far separated on the nuclear sur-
          face, kinetochores (arrows) are al-
          ready connected to the poles via short
          microtubules. X 74,800. (From Heath
          and Greenwood, 1970; reproduced by
          permission of The Cambridge Univer-
          sity Press.)

1975 and Section IV.A.1.), such early and perhaps even continuous connection of chromosomes and spindle pole bodies may be

a general feature of fungal mitosis and one which would explain certain observations of unusual chromosome behavior in
these organisms.

    The time at which a direct connection between kineto-
chores and poles forms would have some bearing on when sister
kinetochores may disjoin. In conventional mitosis, sister
chromosomes remain associated at their centromeric regions
all through preanaphase stages, separating only in preparation for the anaphase poleward movement of sister chromosomes
toward opposite poles (Nicklas, 1971). In fungi such as
Fusarium (Aist and Williams, 1972) and Poria (Setliff, et al.,
1974), this is true also; and pairs of kinetochores are found
right until the initiation of anaphase. However, Heath,
(1974b) found in his examination of serial sections of Thraustotheca that close pairwise association of all chromosomes is
not the case during preanaphase. In as yet unpublished
studies of another Oomycete, Saprolegnia, Heath (personal communication and this volume) has discovered that paired kinetochores are also not observed in early stages of mitosis. As
division begins, a single spindle pole body is connected via
microtubules with several kinetochores. Somewhat later, two
spindle pole bodies and twice as many kinetochores are present, half of the kinetochores being connected to each spindle pole. Yet later, there is an apparent pairwise disposition of kinetochores, the members of each pair having microtubules directed to opposite spindle poles. However, Heath's
observations of the earlier division stages make it clear
that such kinetochore association in Saprolegnia is not
strictly analogous to the sister kinetochore association of
conventional mitosis where replicated chromosomes maintain
sister-sister proximity through premetaphase stages.

    While these tantalizing pieces of information do not al-
low generalization about preanaphase chromosome behavior in
fungi, they certainly do underscore the need for more pene-
trating scrutiny of the important early stages of mitosis.
Do all chromosomes in fungi have a rather persistent connection with the spindle pole body, one that is transformed from
a direct chromosome-spindle pole body connection to a more
indirect interaction via microtubules as mitosis is initiated?
Does the presence or absence of such connections relate to
any other features of mitosis; e.g., is the presence of connections invariably correlated with the nonrandom delivery of
all chromosomes of a given generation to one daughter cell as
described by Rosenberger and Kessel (1968) for Aspergillus
(see Section IV.A.1.)? Is formation of a clear-cut metaphase
plate as is observed in many fungi a signal that earlier
stages of mitosis are entirely conventional?

IV. EVOLUTION OF MITOSIS IN FUNGI

I would like to discuss the unusual features of fungal mitosis from two slightly different points of view. In doing so, I will attempt to deal with two questions: i) Can we learn something about the evolution of mitosis by studying nuclear division in the fungi? and ii) Does our present know-ledge of fungal mitosis have a bearing on fungal phylogeny?

A. Evolution

1. Possible Membrane Functions

In the earliest stages of the evolutionary transition be-tween membrane-mediated and microtubule-mediated chromosome movement, we would expect that there occurred hybrid mecha-nisms where both membrane and microtubular functions contrib-uted in an important way to the mitotic process. The nuclear divisions in dinoflagellates and hypermastigote flagellates discussed above (Section II) are not only examples of the sorts of transitional mechanisms which may have led to typi-cal eukaryotic mitosis but are also an assurance that such hybrid mechanisms survive in present-day organisms. There-fore, although we are able to account for anaphase chromosome movement in many of the fungi in terms of microtubule-mediated mechanisms, we must yet discover if the fact that the nuclear envelope remains at least relatively intact throughout mitosis has any significant influence on chromo-some behavior. As justification for the idea that the simple demonstration of microtubule-mediated chromosome movement in one division stage does not preclude the occurrence of mem-brane-mediated movement in another stage, we need only recall the example of Trichonympha where it has been shown that both of these modes of movement have a role in mitosis (Kubai, 1973).

The intimate association between spindle pole bodies and the nuclear envelope make this an appropriate place to look for signs that the nuclear envelope has a role in fungal mito-sis. Pickett-Heaps (1972) has suggested that the spectrum of spindle pole body-nuclear envelope interactions which charac-terize the various fungal subdivisions are part of an evolu-tionary continuum and reflect a gradual separation of the mi-crotubule organizing centers from nuclear envelope dependence. In fungi where the spindle pole body is a rather simple dif-ferentiation of the nuclear envelope (some Mastigomycotina), attached to the nuclear envelope (mucoralean zygomycetes), or

actually inserted in the nuclear envelope (e.g., Saccharomyces)(for references, see Table VII in Kubai, 1975), behavior of the spindle pole body is interpretable in terms of its
interaction with membrane. For example, the initial separation of spindle pole bodies in Saccharomyces seems to involve
their movement in the plane of the nuclear envelope, a phenomenon that is reminiscent of kinetochore movement in hypermastigotes. In Oomycetes and some Chytridiomycetes, elongation of microtubules may provide the actual motive force for
spindle pole body separation but nevertheless these spindle
pole bodies must move in or on the nuclear envelope. Thus,
in these fungi where the nuclear envelope is directly concerned with spindle pole body behavior, we are able to dis-
cern at least an element of membrane participation in fungal
mitosis.

     There are, in addition, some admittedly tenuous reasons
to think nuclear envelope-spindle pole body interactions
might have a more direct role in chromosome behavior. As I
have already pointed out, there are indications that during
interphase there is an association between chromatin and the
spindle pole body region of the nuclear envelope. Aist and
Williams (1972) propose that this is kinetochoric heterochromatin. Of course, it would not be surprising to find kinetochores near the pole since their poleward movement in the
previous anaphase brought them exactly to this position. However, in conventional interphase, kinetochores can redistrib-
ute in the nucleus (e.g., Moens and Church, 1977). Since
Heath and Greenwood (1970) have demonstrated that Saprolegnia
kinetochores are closely arrayed around the spindle pole body
in early prophase, it is possible that a physical connection
between nuclear envelope and chromatin restrains fungal chromosomes. Again we are reminded of Trichonympha (Kubai, 1973)
and Syndinium (Ris and Kubai, 1974) (see Section II) where
kinetochores remain associated with the nuclear envelope in
interphase. If this is also true in the fungi, chromosomes
would maintain their connection to spindle pole bodies
throughout the cell cycle, during interphase in the form of
chromatin-membrane linkage and in division via kinetochore-
microtubule connection.

     Some of the unusual aspects of chromosome behavior in
the fungi might be directly traceable to an interphase
chromatin-nuclear envelope association. For example, it
seems possible that the poles toward which a given set of kinetochores is oriented might be fixed by a previous association of kinetochore heterochromatin and spindle pole body:
e.g., all kinetochores of a given cellular generation are
associated with the spindle pole body of the interphase nucleus while all newly-formed kinetochores are obliged to asso-

ciate with the newly-formed spindle pole body which appears
in preparation for nuclear division. Such behavior would
place a constraint on chromosome orientation of exactly the
sort required to account for the data of Rosenberger and
Kessel (1968). These authors found that in Aspergillus all
DNA labelled in one generation is distributed together to
daughter nuclei in succeeding generations. This is in direct
contrast to the demonstration for other eukaryotes that labelled centromeric DNA is distributed at random between daughter nuclei (e.g., Mayron and Wise, 1976). There is, moreover,
evidence that DNA replication and spindle pole body duplication are interdependent (Byers and Goetsch, 1973; Poon and
Day, 1976b) a fact which may imply a physical connection between chromosomes and spindle pole bodies.

Unbroken connection between spindle pole body and chromosomes would also help to explain Heath's (1974b) observation
that all kinetochores are not paired at metaphase in Thraustotheca and, indeed, are clearly not paired throughout early
mitosis in Saprolegnia (Heath, personal communication and this
volume). These findings suggest that at least in certain fungi there is something which obviates the need that chromosomes remain joined until anaphase. Precisely the sort of
mechanism discussed above would serve: if appropriate orientation to opposite spindle poles is achieved at the time of
spindle pole body replication, a period of preanaphase chromosome orientation concommitant with spindle formation (i.e.,
prometaphase) is unnecessary.

While the available evidence does not compel us to believe that there is an element of membrane involvement in fungal chromosome behavior, I do believe the fact that some features of fungal mitosis are understandable in these terms is
sufficient grounds to pursue the idea in future investigations.

2. Microtubule Functions

The derivation of classical mitosis from membrane-
mediated genophore distribution such as we know in present-
day prokaryotes would, of course, require the evolution of a
new structural entity, the microtubule. In addition, it
would be necessary that these organelles be designed to participate in a system which can generate force for chromosome
movement. Thus, gradual mitotic evolution may have occurred
in parallel with the evolutionary development of microtubule
functional properties. Discussions of the evolution of mitosis have considered the manifestations of such changing microtubule function which could be expected and have considered examples suggesting that remnants of the "more primitive"

microtubule functions may persist among living organisms (Heath, 1974a; Kubai, 1975).

In considering the evolution of mitosis in fungi, then, a primary question is whether or not the microtubules which compose the fungal spindle seem to function any differently than those participating in classical mitosis. Unfortunately, we have no detailed understanding of how microtubules are actually involved in the force-producing machinery of the eukaryotic spindle (cf. Forer, 1974 and this volume; Inoué and Ritter, 1975; McIntosh, et al., 1975). However, a wealth of purely descriptive information is available for typical mitosis; there are, in addition, some experimental data and a number of hypotheses of the mechanism of mitosis (see Nicklas, 1971; Forer, this volume). It is therefore possible to ask if any of the features of fungal mitosis are in contradiction to our present understanding of classical mitosis.

Demonstration of chromosomal microtubules and of their interaction with the chromosomes at kinetochores or with chromatin as well as the presence of continuous or interpolar microtubules in representatives of all major fungal groups certifies that these spindles are similar to eukaryotic spindles at the most basic structural level. Poleward chromosome movement in both fungal and classical mitosis is likewise a demonstration of basic functional similarities.

Some models of classical mitosis are based upon the close interaction of chromosomal microtubules and the interpolar or continuous microtubules composing overall spindle structure (McIntosh, et al., 1969; Nicklas, 1971) and analysis of microtubule distribution at various stages of nuclear division is one means of evaluating the possible functional role of microtubules (Brinkley and Cartwright, 1971; McIntosh and Landis, 1971). The ultrastructural studies of so-called two-track mitosis (Aist and Williams, 1972; Setliff, et al., 1974) which include the observation that chromosomes and their associated microtubules are arranged around the outside of a central sheaf of continuous microtubules certainly allow us to wonder if close chromosomal microtubule-interpolar microtubule interactions do occur in the fungi. Heath and Heath (1976) also point out that chromosomes are arranged around the periphery of a central group of microtubules in the basidiomycete Uromyces, and although they did analyze microtubule distribution in serial sections, the exact relationships between various classes of microtubules was impossible to define owing to the difficulty of recognizing chromosomal microtubules. There is only one fungus for which the overall pattern of microtubule type and distribution is known (Heath, 1974b). This work concerns Thraustotheca in which chromosomes are not arranged around the periphery of a central spindle. Based upon his findings that intertubule distances are too great to allow microtubule-microtubule interactions via cross-bridging, Heath has argued that models for chromosome movement requiring such interaction cannot account for chromosome movement in these and other fungi in which microtubules are not fairly closely packed. It is, however, difficult to accept these data as evidence against the universality of certain models, as Heath has done, since this conclusion is based upon negative evidence: if an important class of microtubules is disrupted during fixation for electron microscopy, the true microtubule distribution cannot be known and the observed microtubule-microtubule spacing would only reflect fixation artefact. Conversely, closely spaced microtubules in fixed materials may be the result of shrinkage artefacts. Therefore, the simple measurement of microtubule spacing has no real force in helping us decide if the fundamental mechanism of fungal chromosome movement is like that of eukaryotes (especially since it is by no means established that microtubule-microtubule interactions are in fact the basis for anaphase chromosome movement).

3. Chromosome Behavior

Our conception of mitosis in the classical sense includes the recognition that prometaphase chromosome movement generally brings chromosomes to a position equidistant between spindle poles (metakinesis). After all of the chromosomes have assumed this position, we observe the metaphase plate and only then does anaphase proceed, sister chromosomes disjoining and moving to opposite poles (reviews: Schrader, 1953; Mazia, 1961). Although the most comprehensive models of mitosis incorporate an explanation for formation of a distinct metaphase plate (Dietz, 1969; Inoué and Sato, 1967; McIntosh, et al., 1969; Oestergren, 1949; Oestergren, et al., 1960), it is not at all clear from available knowledge that this is a prerequisite for appropriate chromosome distribution. Indeed the manifest failure of many fungi to achieve a plate-like metaphase militates against the idea that this metaphase configuration is a necessary precondition for appropriate chromosome distribution. Nonetheless, the formation of a metaphase plate seems to have enjoyed a selective evolutionary advantage since it is a general condition of conventional mitosis, and it seems unlikely that such precise chromosome behavior could have evolved gratuitously. Could it be that the early mitotic events in fungi which do not display classical division figures differ significantly from the prometaphase events in more usual forms of mitosis? May not aspects of preanaphase chromosome behavior either circumvent the need for metaphase equilibrium or actually prevent its achievement? Thus, we may at least question the evolutionary implications of the fact that a large number of fungi do not exhibit classical metaphase stages. Final interpretation of the fundamental significance of metaphase plate formation or lack thereof will only be possible after we have a thorough understanding of the mechanism and control of prometaphase chromosome movement in both fungi and more typical eukaryotes.

At least one model of chromosome orientation and congression during prometaphase (Oestergren, 1949; Oestergren, et al., 1960) postulates that metaphase is a direct consequence of the production and regulation of chromosome-moving forces during this stage, i.e., that metaphase is a stage of stable equilibrium which is achieved because poleward forces applied to oppositely oriented kinetochores during prometaphase are exactly balanced when a chromosome reaches the midpoint between spindle poles. If the metaphase configuration is such a strong indicator of the strength and directionality of poleward forces operating during prometaphase, the existence of less balanced chromosome configurations in some of the fungi may well signal that control of chromosome movement prior to anaphase is appreciably different from that in conventional mitosis. As I have emphasized in my brief outline of Oestergren's ideas, mitotic chromosome structure typically incorporates the back-to-back placement of the kinetochores of sister chromosomes (for examples and discussion, see Nicklas, 1971). That Heath (personal communication and this volume) has found no sign of such kinetochore disposition in early division of Saprolegnia is suggestive that such chromosome structure is not required for appropriate preanaphase chromosome behavior. If this fact is not explanable in terms of a spindle pole body-chromosome connection as I have suggested above (Section IV.A.1.), it does require explanation in terms of prometaphase chromosome orientation. Fungi which do show the paired configuration of kinetochores throughout prometaphase may then be farther along in the evolution of this stage.

The repeated use of "if", "might", and "possibly" that I have found necessary for this discussion makes it very clear that nothing definite about the evolution of mitosis has emerged from the study of fungal mitosis. I hope, nevertheless, that I have given the impression that an understanding of fungal mitosis will make important contributions to our knowledge of mitosis in general, if not in elucidating its evolutionary development, then by better defining the essential character of this all-important process.

B. Phylogeny

1. Relationships Within the Major Subdivisions of Fungi

Although variations in the mitotic characteristics of various fungi allow some speculation regarding their evolutionary significance in fairly broad terms, more detailed comparisons must be made in order to understand what these features imply about fungal phylogeny. It is unfortunate that the most fundamental aspect of mitosis, chromosome behavior, is so poorly understood in the fungi as a whole. Clearly, for example, the contrast between two-track division figures so common in Ascomycotina and Basidiomycotina and the more conventional figures seen in other Ascomycotina and Mastigomycotina must be understood in a phylogenetic context. Until more information is available, especially regarding the mechanics of chromosome orientation and other aspects of preanaphase chromosome behavior, this dichotomy will remain unexplained. It is impossible, therefore, to suggest even in a general way how the various fungi might be related from the viewpoint of chromosome behavior. Among the Mastigomycotina, there is a fairly obvious distinction between the Oomycetes where clear-cut metaphase plates are not observed and the Chytridiomycetes and Hyphochytridiomycetes where conventional metaphase configurations seem to be more generally the rule (see Table in Heath, 1975). This lends credence to other indications that Oomycetes occupy a special position in fungal phylogeny (e.g., Klein and Cronquist, 1967, pp. 254-255) and that they indeed may even constitute a separately derived taxon. It is well to remember, however, that even in the Oomycete Thraustotheca chromosomes come to occupy a more-or-less equatorial position (Heath, 1974b). Thus, the appearance of plate-like versus more scattered "metaphases" among Mastigomycotina may be a qualitative difference rather than a signal of profoundly different premetaphase chromosome behavior in various groups.

On the basis of present knowledge, then, discussions of fungal phylogeny must be limited to the more peripheral aspects of mitosis such as spindle pole body structure and behavior and some information on spindle formation.

a. Mastigomycotina. The Mastigomycotina are distinguished from the remainder of Eumycota by the occurrence of centrioles near the spindle poles. Pickett-Heaps (1971) has presented the argument that a centriole is not an integral component of the mitotic apparatus but is rather a reflection of the occurrence of flagellated stages in the life cycle at

which time centrioles function as basal bodies in flagella
formation. This certainly holds true for Mastigomycotina.
Therefore, I will not comment on the phylogenetic significance of centriole behavior, a subject which has been dealt
with by Heath (1974c, 1975). For our purposes, the functional equivalents of the differentiated spindle pole bodies
found in the remainder of the Eumycota are in the Mastigomycotina identifiable as those differentiations which occur at
the polar termini of spindle microtubules. In certain cases
(see below), the spindle pole body (= microtubule organizing
center) may be closely associated with the centriole but is
not to be identified with this organelle.

In the second grouping of Mastigomycotina, it is possible to differentiate two broadly different mitotic types. In one group we
may place the order Chytridiales, e.g., Entophlyctus (Powell,
1975) and Phlyctochytrium (McNitt, 1973) as well as the order
Harpochytridiales as represented by Harpochytrium (Whisler
and Travland, 1973). These orders of the Chytridiomycetes
are distinguished by opening of the nuclear envelope during
mitosis (polar fenestrae), a feature that seems to be correlated with an exclusively extranuclear spindle pole body.
That is, spindle growth is initiated extranuclearly, microtubules appearing in association with vaguely differentiable
microtubule organizing centers (= spindle pole bodies) positioned close to centrioles. After these spindle pole bodies
have taken positions at opposite poles of the nucleus, gaps
appear in the polar regions of the nuclear envelope and microtubules invade the nucleoplasm to form the intranuclear
spindle.

In the second grouping of Mastigomycotina, we include
Allomyces (Olson, 1974), Catenaria (Fuller and Calhoun, 1968;
Ichida and Fuller, 1968), Blastocladiella (Lessie and Lovett,
1968), Saprolegnia (Heath and Greenwood, 1968, 1970), and
Thraustotheca (Heath, 1974b), representatives of the Oomycetes (Saprolegniales) and Chytridiomycetes (Blastocladiales).
In these genera, the spindle pole body is intranuclear and is
identifiable as a layer of electron dense, rather amorphous
material closely appressed to the inner aspect of the nuclear
envelope. Often, the nuclear membranes in this region are
thickened and/or more electron dense. While actual spindle
formation is intranuclear, extranuclear microtubules are
found also, and these center on microtubule organizing centers close to centrioles.

In so far as Chytridiales and Saprolegniales display divergent mitotic types, there is some accord with the ideas of
Scherffel (quoted by Sparrow, 1960, pp. 4-6) who views the
aquatic phycomycetes as forming two phylogenetic series:
"Chytridineen" versus "Saprolegniineen". Having intact nu-

clear envelopes, the Blastocladiales, which I have treated as
Chytridiomycetes in accordance with Ainsworth (1971), clearly
do not conform. Since other features (Fuller and Calhoun,
1968; Chong and Barr, 1974) also suggest that Blastocladiales
should be separated from the rest of the chytrids, some taxonomic revision may accomodate these discrepencies. It is to
be regretted that Chytridiomycetes of the order Monoblephari-
dales, considered by Sparrow (1960) to be closely related to
the Blastocladiales, have not been examined for any light they
may shed on this question.

Fuller (1976) has suggested that the distinction between the mitotic types found in the Mastigomycotina can be
understood in terms of the number and positioning of microtubule organizing centers functioning during mitosis. In the
Blastocladiales and the Oomycetes, intranuclear as well as
extranuclear microtubule organizing centers are involved during mitosis, the former concerned with elaboration of the intranuclear spindle and the latter having a role in formation
of cytoplasmic microtubules. On this interpretation, evolution of the form of mitosis found in Chytridiales and Harpochytridiales resulted through the loss of the intranuclear
microtubule organizing center with concommitant transferance
of all responsibility for microtubule formation to the cytoplasmic center. That is, deprived of their intranuclear microtubule organizing centers, these organisms would require
the opening of the nuclear envelope to allow formation of the
intranuclear spindle under the direction of extranuclear spindle pole bodies.

If any of the Mastigomycotina are to be considered as
closely related to mucoralean zygomycetes and/or the Ascomycotina with their predominantly closed division figures, then
either the Oomycetes or the Blastocladiales have the requi-
site characteristics, closed mitosis plus retention of both
extranuclear and intranuclear spindle pole bodies. It is difficult to find any mitotic features which dictate a choice
between these groups but other parameters certainly do place
the Oomycetes off the main line of fungal evolution (e.g.,
LéJohn, 1974) and it is possible on these grounds to consider
the Blastocladiales as the reasonable progenitor of higher
Eumycota.

Based upon his study of enzymatic controls operating on
glutamic dehydrogenases, LéJohn (1974) suggested that the
Hyphochytridiomycetes may occupy a key position in defining
the relationships between Blastocladiales and Oomycetes; although these data support the idea that Mucorales are more
closely related to Blastocladiales while Oomycetes are closer
to the Hyphochytridiomycetes, they also indicate that Blastocladiales and Hyphochytridiomycetes may have had a common an-

cestor. It is unfortunate that the only published evidence on mitosis in the Hyphochytridiomycetes is Fuller and Reichle's (1965) single micrograph of mitotic Rhizidiomyces. This evidence might be taken as indicative of completely closed mitosis, a character Hyphochytridiomycetes would then share in common with Oomycetes and Blastocladiales. Recently, however, Fuller (personal communication) has made more extensive observations of mitosis in Rhizidiomyces and opening of the nuclear envelope at polar regions is clearly demonstrable. Thus, the condition of the nuclear envelope during mitosis does not support the idea that these organisms are a bridge between Blastocladiales and Oomycetes. Rather, in this regard the Hyphochytridiomycetes are more like Chytridiomycetes of the orders Harpochytridiales and Chytridiales.

    b. Zygomycotina. A completely closed nuclear envelope is present in some Zygomycetes of the order Mucorales. Since all of the genera which have been studied (Phycomyces: Franke and Reau, 1973; Mucor: McCully and Robinow, 1973; Pilobolus: Bland and Lunney, 1975) are members of a single family, Mucoraceae, it is not really possible to generalize at the next higher taxonomic rank. Spindle pole bodies found in the Mucoraceae are in the form of electron dense knobs which are clearly intranuclear and closely applied to the nuclear envelope. There is no indication that extranuclear microtubule organizing centers play a role during these mitoses and if these Zygomycetes are derived from more primitive Mastigomycotina-like progenitors, their evolution will have involved loss of the extranuclear mitosis-related microtubule organizing center. On these grounds it is impossible to think of the Mucoraceae as being a link between Mastigomycotina and the Ascomycotina where the diverse functions attributed to spindle pole bodies include a role in the formation of cytoplasmic (astral) microtubules (e.g., Byers and Goetsch, 1973; Beckett and Crawford, 1970).

    Among the order Entomophthorales, studies of mitotic ultrastructure are available for a single representative of two of the three families. In Basidiobolus (family Basidiobolaceae) extensive (Gull and Trinci, 1974) or total (Tanaka, 1970; Sun and Bowen, 1972) dispersal of the nuclear envelope occurs. In contrast, Ancylistes (family Entomophthoraceae) has a persistent nuclear envelope during mitosis (Moorman, 1976). Tanaka (1970) indicates that other members of the family Entomophthoraceae also have intranuclear mitoses (based on light microscopy). Whether these closed divisions are a secondary modification required by the coenocytic hyphae of these genera [cf. Physarum where the occurrence of

closed mitoses in coenocytic plasmodia is interpretable on
these grounds (Pickett-Heaps, 1974)] or are indicative of a
more basic dichotomy between families of the Entomophthorales
cannot be resolved until more representatives have been examined. Similarly, the ultrastructure of the spindle pole
bodies differs between these Entomophthorales. In Ancylistes,
there is an extranuclear mass of electron dense material on
the outer membrane of the nuclear envelope plus a less dense
layer lining the inner membrane of the envelope. In the published micrographs the nuclear envelope appears to be continuous between these two portions of the spindle pole body, but
serial section evidence is not presented. In Basidiobolus,
an apparently cylindrical body occurs at the polar regions
of mitotic nuclei, but spindle microtubules do not seem to
focus on this differentiation so it remains problematical
whether this corresponds to the spindle pole body of other
fungi (see discussion in Heath, this volume).

In summary, with our present knowledge it is difficult
to generalize about mitotic characters in the Zygomycotina.
It seems sensible for the time being to treat these organisms
as a side-branch on the main evolutionary line of fungal mitosis.

c. Ascomycotina. Throughout the subdivision, a completely intact nuclear envelope during mitosis seems to be a constant, the only reported exception being Xylaria (Schrantz,
1970) where the quality of fixation was less than ideal. Because representatives of four of Ainsworth's (1971) six ascomycete classes are represented (see Table II in Kubai, 1975),
we presently have no grounds for doubting that closed mitosis
characterizes the Ascomycotina.

Spindle pole bodies are closely associated with the
nuclear envelope of Ascomycotina and a fair amount of variation in the detailed ultrastructure of this organelle is evident. In view of the fact that taxonomic groupings within
the Ascomycotina remain very tentative (Ainsworth, 1971, pp.
41-42 for several possible taxonomic schemes) it is perhaps
not surprising that strict correlation between type of spindle pole body and taxonomic position is not possible. For
example, among the Hemiascomycetes, all of the genera which
have been examined are assigned to a single family; but the
described spindle pole bodies are very different. Simple
membrane thickening (Wickerhamia: Rooney and Moens, 1973),
and multilaminar plaques, either extranuclear (Schizosaccharomyces: McCully and Robinow, 1971; but see the micrographs of
this genus in Heath, this volume) or inserted in the nuclear
envelope (Saccharomyces: Robinow and Marak, 1966; Moens, 1971;
Rapport, 1971; Moens and Rapport, 1971; Peterson, et al.,

1972) are described. Similarly, in the class Pyrenomycetes, the four genera which have been examined all belong to the order Sphaeriales. Three quite different spindle pole body types were found: extranuclear plaque (Xylaria: Schrantz, 1970; Xylosphaera: Beckett and Crawford, 1970), extranuclear L-shaped body (Podospora: Zickler, 1970), and extranuclear diglobular structure (Neurospora: Van Winkle, et al., 1971). The order Sphaeriales is a broadly defined grouping (Mueller and von Arx, 1973) so it might be hoped that spindle pole body structure would show correlations at lower taxonomic levels. However, even within the family Sordariaceae, Neurospora and Podospora have different spindle pole body structure. Among the Discomycetes, representatives of different families of the order Pezizales have been included in ultra-structural investigations and in both extranuclear plaque-like spindle pole bodies are present (Ascobolus: Zickler, 1970; Wells, 1970; Pustularia: Schrantz, 1967, 1970).

          d. Basidiomycotina. Information on mitosis in this subdivision is sparce; nuclear division in seven of the nine orders of Hymenomycetes defined by Ainsworth (1971) have been neglected. Consequently, our knowledge is restricted to the Aphyllophorales (Girbardt, 1968, 1971; Setliff, et al., 1974) and Agaricales (Lu, 1967; Lerbs and Thielke, 1969; Lerbs, 1971; Raju and Lu, 1973; Motta, 1969; Gull and Newsam, 1975). In general, spindle pole bodies are described as cytoplasmic organelles having a more-or-less spherical or globular structure and their proximity to the nuclear membrane is noted. When some attempt is made to compare spindle pole body structure at various stages in the cell cycle, there are indications that cyclical variations in the form of this organelle occur (Girbardt, 1968, 1971; Raju and Lu, 1973). During interphase, the spindle pole body can have a diglobular conformation while during nuclear division monoglobular spindle pole bodies are found at each pole. Such findings prompt the suggestion that the diglobular element is a spindle pole body which has duplicated in preparation for mitosis.

          A most striking characteristic of mitosis in the Hymenomycetes is opening of the nuclear envelope in the spindle pole regions during mitosis. It seems such dispersal of the nuclear envelope is a necessary modification to allow the extranuclear microtubule organizing centers (= spindle pole bodies) to function in building the intranuclear spindle.

          Similarly, in some lower Basidiomycetes, specifically Teliomycetes of the order Ustilaganales (Ustilago: Poon and Day, 1976a, 1976b) and in the yeast forms (Aessosporon, Leucosporidium and Rhodosporidium: McCully and Robinow, 1972a, 1972b) opening of the nuclear envelope occurs, but this seems to be a more complicated process than the simple polar fenestration seen in the Hymenomycetes mentioned above. The spindle pole body moves through these openings into the nucleoplasm proper and growth of microtubules between spindle pole bodies then takes place.

Thus, for a number of the Basidiomycotina, it seems that we may make the generalization that the nuclear envelope does not remain completely intact during mitosis.

Studies of mitosis in two Teliomycetes of the order Uredinales (<u>Puccinia</u>: Harder 1976a, 1976b; <u>Uromyces</u>: Heath and Heath, 1976) have given a somewhat different picture for these lower Basidiomycotina. Interphase spindle pole bodies are multipartite cytoplasmic structures closely appressed to the nuclear envelope. They consist of two more-or-less discoidal elements connected by (<u>Uromyces</u>) or lying on (<u>Puccinia</u>) an amorphous osmiophilic layer. In the course of mitosis, spindle pole body structure changes so that a simpler discoidal structure is found at each pole. Each of the mitotic spindle pole bodies is apparently derived from one of the discoidal elements of the interphase spindle pole body (Heath and Heath, 1976). The relationship between spindle pole body and nuclear envelope is also transformed during mitosis so that the polar organelle comes to lie within the plane of the nuclear envelope, i.e., occupying a pore-like opening of the nuclear envelope. However, the discontinuity of the nuclear envelope in the polar regions is just sufficient to encompass the spindle pole body; and, therefore, disruption of the nuclear envelope is nowhere near so extensive as it is in the higher Basidiomycotina or the Ustilaginales.

Uredinales thus seem to be distinguished from the remainder of the Basidiomycotina in that i) they have a more plaque-like spindle pole body than is usual in Basidiomycotina and ii) there is a lesser degree of opening of the nuclear envelope. These features are certainly compatible with the generally held view that Uredinales are primitive among the Basidiomycotina. Further, they provide important criteria upon which to judge the evolutionary link between Ascomycotina and Basidiomycotina (Heath and Heath, 1976 and see discussion below).

2. Relationships Between the Major Subdivisions of Fungi

In Hemiascomycetes such as <u>Saccharomyces</u>, we find justification for Fuller's (1976) idea that fusion of intranuclear and extranuclear microtubule organizing centers found in a Blastocladiales-like ancestor may have occurred during the evolution of Ascomycotina. In <u>Saccharomyces</u>, the spindle pole body is neither intranuclear nor extranuclear but occupies a position between these extremes: a multilayered disc plugs a pore-like opening in the nuclear envelope and has a role in cytoplasmic as well as intranuclear microtubule elaboration (Byers and Goetsch, 1973). Since some lower Basidiomycotina, the Uredinales, have rather similar spindle pole bodies which are also involved in both nuclear and cytoplasmic microtubule formation (Heath and Heath, 1976), Fuller's conception may now be extended to include Teliomycetes as having "fused" spindle pole bodies. Thus, it is possible to view Ascomycotina and Uredinalean Teliomycetes as intermediates between Mastigomycotina of the Blastocladiales type where separate intra- and extranuclear microtubule organizing centers participate in mitosis and the higher Basidiomycotina where only extranuclear spindle pole bodies are present.

Heath and Heath (1976) maintain that mitosis in Uredinales provides some key evidence for understanding the phylogenetic relationship between Ascomycotina and Basidiomycotina. They point out that the plaque-like spindle pole body structure and the limited opening of the nuclear envelope in polar regions is more similar to these features in Ascomycotina than they are to Basidiomycotina. In fact, it seems possible to draw the closest parallels with Hemiascomycetes such as Saccharomyces where during mitosis the plaque-like spindle pole bodies also occur in pore-like openings of the nuclear envelope. In Saccharomyces, such relationship with the nuclear envelope is best-documented only for dividing cells (Peterson, et al., 1972; Byers and Goetsch, 1973), so it is not known if interphase spindle pole bodies might be extranuclear as they are in Uredinales. In Schizosaccharomyces, however, there is now evidence that an extranuclear spindle pole body is converted during mitosis into one that plugs a pore-like opening of the nuclear envelope (see figures in Heath, this volume). The parallel with Uredinalean Teliomycetes is thus striking.

Savile (1968) concluded that the ascomycete-basidiomycete connection was to be found among the Hemiascomycetes. But, he suggested specifically that primitive Hemiascomycete genera belonging to the orders Taphrinales and Protomycetales are the most likely progenitors of both higher Ascomycotina and Basidiomycotina (the Hemiascomycetes discussed above belong to another order, Endomycetales, and are not considered primitive). Therefore, it will be most interesting to see if studies of mitosis in genera such as Taphrina and Mixia advance the argument for a Hemiascomycete-Teliomycete relationship. At present, we have only the observations of interphase Taphrina (Syrop and Beckett, 1976) where the extranuclear bipartite spindle pole body is certainly reminiscent of similar structure in Uredinales.

3. Ancestors of the Fungi

Basic understanding of fungal mitosis and fungal phylogeny will ultimately rest upon tracing the more distant relatives of the Eumycota. There is a long history of specula-
tive discussion attempting to establish a relationship between fungi and various algal or protozoan groups (e.g.,
Bessey, 1950) but in the end there is no concensus among mycologists as to where fungi fit in the biological world.
Whittaker's (1969) taxonomy which reserves a separate kingdom
for the fungi epitomizes this dilemma, as does Fuller's
(1976) treatment of fungal evolution in several separate
lines, all emanating from an unspecified "flagellate" ancestor.

If the study of fungal mitosis can have value in elucidating relationships among the fungi (and there is as yet no
good reason to doubt this), it seems that comparison of fungal mitosis with mitosis in the rest of Whittaker's (1969)
three eukaryotic kingdoms might be useful in attempting to
identify the ancestors of fungi. Since the animal and plant
kingdoms comprise organisms where conventional mitosis is the
rule and we have already asked some questions about the relationship between fungal mitosis and conventional mitosis, we
will now direct our attention to the Protista and certain
lower plants where examples of somewhat atypical mitosis are
to be found.

a. Mitosis in Presumptive Ancestors of Fungi. During
the past century, speculations regarding fungal ancestry focused on two groups of algae, the yellow-green Xanthophyceae
and the red Rhodophyceae. In his chapter on fungal phylogeny, Bessey (1950) reviewed similarities of morphology and
life cycle which hint at a relationship between these algae
and certain fungi.

Modern biochemical evidence (Klein and Cronquist, 1967,
p. 254 and ff.) reinforces the more traditional structural indications that lower fungi (Mastigomycotina) were derived
from the biflagellated Xanthophyceae. Depending on the value
attributed to the characters of flagellar type, cell wall
chemistry, and lysine biosynthetic capabilities, alternative
origins of Mastigomycotina from Xanthophyceae can be considered: i) a monophyletic derivation of all Mastigomycotina
from Xanthophyceae or ii) diphyletic origin of the various
Mastigomycotina, Oomycetes (and perhaps Hyphochytridiomycetes) descending from Xanthophyceae and Chytridiomycetes
arising from euglenoid and/or dinoflagellate ancestors.

The Xanthophycean genus <u>Vaucheria</u> has often been singled out because it displays an array of structural characters which could be expected in a progenitor of the lower fungi (for references, see Bessey, 1950). In particular, the resemblance between the spermatozooid of these algae and the zoospores of Oomycetes taken together with several biochemical traits are grounds for thinking the two groups are related (Klein and Cronquist, 1967). The ultrastructure of mitosis in <u>Vaucheria</u> was examined by Ott and Brown (1972); retention of a completely closed nuclear envelope during mitosis is certainly an Oomycete-like feature and one that fits in the list of similarities between Oomycetes and Xanthophyceae. However, the mitotic parallels are not strict. Features which are highly characteristic of the Oomycetes (Heath, 1975), end-to-end centriole orientation and the construction of an intranuclear spindle even as centrioles migrate to opposite poles of the nucleus, are not found in <u>Vaucheria</u>. In addition, the interzonal microtubules in <u>Vaucheria</u> are persistent until telophase when they are pinched off from the daughter nuclei by an invagination of the inner nuclear membrane. This behavior is not found in Oomycetes. On the whole, then, mitosis in <u>Vaucheria</u> can be more closely compared to mitosis in other Mastigomycotina, i.e., those of the order Blastocladiales among the Chytridiomycetes. In <u>Catenaria</u>, for example, centrioles lie more-or-less orthogonally to each other and they migrate to opposite spindle poles before intranuclear spindle formation. Also, the interzonal spindle persists until telophase (Ichida and Fuller, 1968).

Thus, although mitosis in <u>Vaucheria</u> does support a general relationship between Xanthophyceae and Mastigomycotina, any polyphyletic schemes which postulate that Oomycetes arose from Xanthophyceae while other classes of flagellated fungi were separately derived are troublesome. Mitotic criteria suggest a closer affinity between Blastocladiales and Xanthophyceae. It seems then that we are compelled to add the weight of evidence from mitosis to the structural and biochemical data which point to a monophyletic origin of the lower fungi from Xanthophyceae.

Certain of the golden-yellow algae, the Chrysophyceae, are related to the Xanthophyceae and further have the flagellar characteristics to be expected of Chytridiomycete ancestors (Fritsch, 1935, p. 642; Klein and Cronquist, 1967). But, just as Klein and Cronquist find it difficult to support a chrysophyte-chytrid relationship on grounds of differing lysine biosynthetic pathways, the occurrence of open, apparently classical mitosis in <u>Ochromonas</u> (Slankis and Gibbs, 1972; Bouck and Brown, 1973) and <u>Prymnesium</u> (Manton, 1964)

eliminates these particular Chrysomonads from consideration
as fungal relatives. However, it is specifically another
groups of Chrysomonads, the uniflagellate Chromulinales,
which have the closest affinities with Chytridiomycetes
(Klein and Cronquist, 1967, p 251) and mitotic ultrastructure
in members of this group ought to be examined for any light
it may shed on this question.

The cytology of nuclear division in euglenoids and dinoflagellates was one of the criteria which led Klein and Cronquist (1967) to think these algae might be ancestral to at
least some lower fungi. With ultrastructural evidence which
became available only after that proposal was made, it is
clear that the mode of nuclear division in the dinoflagellates (see Section II) is so dissimilar to fungal mitosis
that no close relationship can be admitted. As for the euglenoid flagellates, electron microscopy demonstrates closed
nuclear divisions with intranuclear microtubules (Leedale,
1968, 1970) and so their relationship to the fungi is not so
facilely denied. But, since the microtubules in euglenoid
mitosis are simply arranged in several parallel sheaves (i.e.,
they do not compose themselves into a coherent spindle-like
structure focused at distinctive polar structures, as is the
case for fungi), a close affinity between fungi and euglenoid
flagellates is not apparent. Again, the study of mitotic ultrastructure in purported ancestors of Mastigomycotina weak-
ens the argument for polyphyletic derivation of the lower
fungi, and we are left with the Xanthophyceae as possible progenitors of the entire group.

Since the last century, there has been some suspicion
that fungi as a whole constitute a polyphyletic assemblage,
i.e., that the flagellated Mastigomycotina may not be considered as progenitors of the aflagellate Ascomycotina and
Basidiomycotina. In this connection, the Rhodophyceae, organisms which never possess flagella, were brought into discussions of fungal phylogeny. Because similarities of life
cycle and reproductive structures are perceptible in fungi
and red algae, several earlier workers proposed that higher
fungi stem directly from the Rhodophyceae (see Bessey, 1950,
for references and outline of the arguments). However, since
convergent evolution could easily account for these similarities (Savile, 1968), the idea has found little favor among
mycologists. Nonetheless, renewed support has recently been
marshaled for this concept, especially by Chadefaud (1975),
Demoulin (1974), Denison and Carol (1966), and Kohlmeyer
(1975). Chadefaud, for example, feels that even if the various similarities between red algae and ascomycetes result
from convergent evolution, these organisms must have evolved
from a common ancestor which conferred similar evolutionary

potentials on the two groups.

The ultrastructure of mitosis in the red algae has given no cause for outright rejection of these theories; in fact, one investigator claims that ascomycete and Rhodophycean mitosis are quite comparable (Peyrière, 1971). During mitosis in Griffithsia (Peyrière, 1971), the nuclear envelope remains intact but polar areas are distinguished by the absence of nuclear pores. In Membranoptera (McDonald, 1972), in contrast, limited opening of the nuclear envelope at spindle poles occurs during mitosis while in Polysiphonia studied by Scott (personal communication) polar areas of the nuclear envelope are subject to extensive invagination during the course of mitosis. A polar organelle, described as an electron-dense, squat cylindrical structure, lies at the spindle poles (Peyrière, 1971; McDonald, 1972) where it can be found at all division stages (Scott, personal communication). In Membranoptera and Griffithsia, the polar organelle is not closely applied to the nuclear envelope but in the species examined by Scott (personal communication) an intimate junction between nuclear envelope and polar organelles is observed.

Thus, red algae and higher fungi do display common mitotic features insofar as an intranuclear spindle is focused at spindle-pole-body-like organelles. Further division-related similarities between certain fungi and red algae include: i) persistence of the interzonal spindle at telophase in the red algal genera examined by Scott; this is similar to the persistent midpiece of certain Mastigomycotina (see Table in Heath, 1975) and ii) the occurrence of both meiotic divisions within a single nuclear envelope (Yamanouchi, 1906; Scott and Thomas, 1975) in red algae, a meiotic peculiarity that is also noted in the Hemiascomycetes Saccharomyces (Moens, 1971; Moens and Rapport, 1971) and Wickerhamia (Rooney and Moens, 1973) and in the Oomycete Saprolegnia (Howard and Moore, 1970).

With our present knowledge, it is possible to say that the features of mitosis in Xanthophyceae and Rhodophyceae are at least compatible with the notion that fungi had such algal ancestors. However, much more detailed knowledge will be necessary before mitotic characters can be used for a truly critical evaluation of the relatedness of these organisms. In Vaucheria, for example, Ott and Brown's (1972) Figures 20 and 21 suggest that chromosomal microtubules are arrayed around the periphery of a central sheaf of interpolar microtubules. Since Ascomycotina and Basidiomycotina often display such peripheral chromosome placement, it is possible to consider this another fungus-like character of the Xanthophycean Vaucheria. But, is a similar chromosome-spindle in-

teraction ever seen among the Mastigomycotina to which the
Xanthophyceae are presumably most directly related? Similarly, we must await some study of chromosome behavior in red
algae in order to know if these groups have the complete con-
stellation of mitotic characters which would strongly justify
the phylogenetic linkage of fungi and Rhodophyceae.

      b. **Organisms with Fungus-Like Mitoses**. Recently, Olive
(1975) presented a revised taxonomy based upon Whittaker's
(1969) five-kingdom system in which he groups mycetozoans and
their associates as a distinct phylum in the Protista. Without denying any of Olive's arguments for removing the mycetozoans from among the fungi where they have more traditionally
been included, it must be admitted that the more-or-less
closed divisions in dictyostelid mycetozoans which have recently been studied in some detail (<u>Polysphondelium</u>: Roos,
1975; <u>Dictyostelium</u>: Moens, 1976) have the peculiar combina-
tion of characters that one would hope to find in the proto-
type of fungal division. In these species, the spindle pole
body is extranuclear but opening of polar fenestrae allows
formation of an intranuclear spindle. However, this is not
accomplished as it is in Mastigomycotina such as <u>Phlycto-
chytrium</u>, for example, where spindle pole bodies complete
their separation to opposite poles of the nucleus before fenestrae ·appear (McNitt, 1973). Although information on prophase stages is not extensive, Roos presents serial sections
of one prophase nucleus that show a rudimentary intranuclear
central spindle formed by microtubules connecting the two
spindle pole bodies. At this time, spindle pole bodies are
not far separated, and thus the spindle has an extremely eccentric position within the nucleus. Elongation of the central spindle proceeds and brings the spindle to the interior
of the nucleus. Such eccentric spindle development is clearly reminiscent of the course of spindle formation in the
Oomycetes, mucoralean zygomycetes and Hemiascomycetes. When
fully formed, the spindle is composed of a rather coherent
bundle of interpolar microtubules. Because well-differentiated kinetochores are clearly identifiable, it is easy to
recognize that in both <u>Polysphondylium</u> and <u>Dictyostelium</u> the
chromosomes are arranged around the periphery of the central
spindle, a fact which is beautifully illustrated by Moens in
serial sections included in his Figure 5. This is exactly
the sort of peripheral chromosome arrangement found among the
higher fungi, i.e., the so-called two-track mitoses discussed
earlier. Thus, the dictyostelids have almost precisely the
array of mitotic features that one might seek in a "zygoasco-
basidiomycete" progenitor (see Fuller, 1976).

      Looking further afield, it is possible to find other ex-

amples of mitosis that are strongly evocative of fungal nuclear division. The best of these occur in the radiolaria
and their allies and in the sporozoa. Hollande, et al.
(1969) published micrographs of mitosis in the radiolarian
Collozoom which are essentially the idealized representation
of a fungal mitotic figure. Spindle pole bodies consist of
a layer of pronounced electron density appressed over an extended area of the inner nuclear membrane and a sheaf of extremely densely packed microtubules spans the nucleus from
pole to pole. Although these authors did not document clearly the stages of spindle formation, it appears that the earlier stages of division have an eccentric spindle because
microtubules connect two spindle pole bodies while they are
moving apart on the nuclear surface. Chromosomes are
equipped with highly differentiated layered kinetochores and
with their associated microtubules they are arranged around
the periphery of the central spindle. As mitosis proceeds,
chromosomal microtubules shorten and chromosomes move from a
scattered but more-or-less equatorial position toward the
poles.

More recently, Febvre (1977) examined mitosis in the
sporogenetic stages of a number of Acantharia, protozoa which
are closely related to the radiolaria. Again, the nuclear
envelope was found to remain completely intact. Although
polar differentiation of the nuclear envelope is not so well-
developed as in Collozoom, a thin plaque of material appressed to the inner membrane is recognizable at the focus of
spindle microtubules, a formation which Febvre identifies as
the spindle pole body. Early development of the spindle is
well-demonstrated in the Acantharia. Spindle formation is
initiated when microtubules of two half spindles radiate from
adjacent areas of the nuclear envelope (spindle pole bodies)
(Fig. 12). The spindle pole bodies then move apart on the
nuclear envelope, bringing the half spindles into opposition
and thus forming an eccentric bipolar spindle (Fig. 13).

Thus, in radiolaria and their allies parallels with fungal mitosis are evident in i) the overall progress of central
spindle formation between membrane-associated spindle pole
bodies, a feature of Oomycete, Zygomycotina and Hemiascomycete mitosis and in ii) arrangement of chromosomes around the
periphery of the central spindle, the sort of chromosome disposition found in two-track mitoses of Ascomycotina and
Basidiomycotina.

Since the radiolaria may be considered as members of the
subphylum Sarcomastigophora (Honigberg, 1967), a taxon
erected to stress the affinities between Mastigophora (flagellates) and Sarcodina (e.g., amoeba), the discovery of such
striking similarities in fungal and radiolarian mitoses is

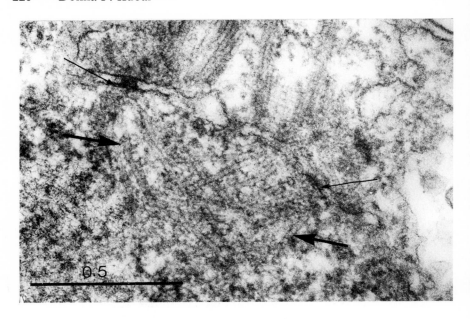

Fig. 12. Acantharia. Spindle microtubules
(larger arrows) first appear in adjacent half spindles radiating from
areas on the nuclear envelope (spindle pole bodies) (smaller arrows).
X 81,000. (From Febvre, 1977; reproduced by permission of Academic
Press, Inc.)

suggestive that fungal mitosis had its antecedents in some
flagellate ancestor.

Although sporozoa are not close relatives of flagellates, there are grounds for tracing a distant relationship
between these protozoa (Grassé, 1952). Therefore, the mitotic resemblance between sporozoa and fungi may also owe its
origin to a flagellate ancestor.

In the sporozoa, mitosis is intranuclear and characteristic organelles occur at the spindle poles (Fig. 14) (e.g.,
Kelley and Hammond, 1972; Dubremetz, 1973; Aikawa, et al.,
1972). Schrével, et al. (1977) have recently given an account of mitosis in Plasmodium that almost exactly duplicates
the course of events in Saccharomyces. Plaque-like spindle
pole bodies are inserted in pore-like openings of the nuclear
envelope and in preparation for mitosis these duplicate and
move apart on the nuclear envelope (Fig. 15). When spindle
pole bodies are yet side-by-side, a fan-like array of micro-

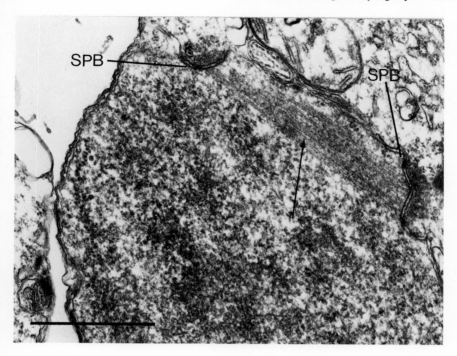

Fig. 13. Acantharia. When the spindle is
         fully elaborated, a sheaf of mi-
         crotubules (arrow) connects two
         separated spindle pole bodies
         (SPB). Note eccentric position
         of the spindle within the nu-
         cleus. X 33,000. (Courtesy of
         J. Febvre.)

tubules emanates from each of them, and final formation of
the spindle results when these two groups of microtubules
interact to form a complete interpolar bundle. Schrével and
his colleagues enjoyed an advantage not had by students of
ascomycete mitosis in that the kinetochores of these sporozoa
are well-differentiated disc-like structures, and so they
were able to follow the behavior of kinetochores through division. Kinetochores connect to the spindle pole bodies via
microtubules at the earliest recognizable division stages,
i.e., when two spindle pole bodies are lying next to each
other on the nuclear envelope and even before spindle structure is completed. Later, kinetochores are found in a rather
equatorial position, but Schrével, et al. report that they
found no evidence that kinetochores are paired at this meta-

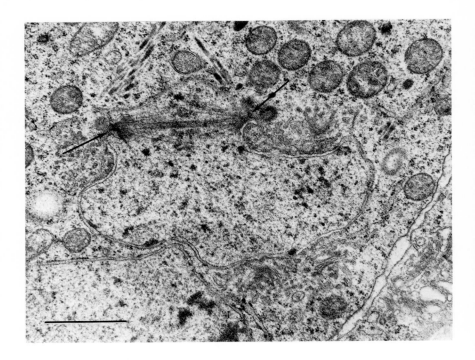

Fig. 14. *Eimeria* ninakohlyakimovae. Centro-
cones (=spindle pole bodies) (arrows)
lie at the foci of the intranuclear
spindle. Note eccentric spindle position. X 21,500. (From Kelley and
Hammond, 1972; reproduced by permission of Springer-Verlag.)

phase-like stage. Thus, chromosome behavior duplicates in
two respects the sort of chromosome behavior Heath (personal
communication) has found in Saprolegnia.
    Let me be the first to disclaim that the simple demon-
stration of mitotic similarities between fungi and Protista
such as the radiolaria and sporozoa should be taken as an
indication that these protists are closely related to fungi.
It is easy to imagine that convergent evolution is responsi-
ble for the resemblances. But, an understanding of how such
similar mitoses evolved in each of these separate groups of
organisms should help us untangle the evolution of mitosis.
Over the long haul, if we are able to comprehend mitosis from
an evolutionary perspective, the phylogenetic significance of
diverse features of fungal mitosis will be more easily evaluated. Or, conversely, if we can understand the basic phylo-

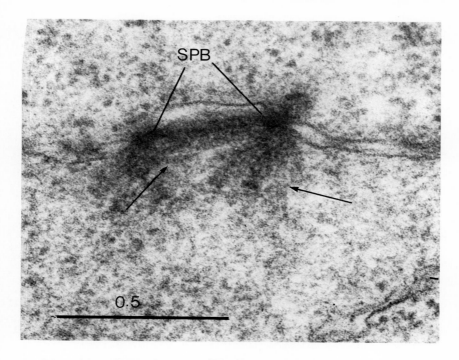

Fig. 15. <u>Plasmodium</u> <u>b</u>. <u>berghei</u>. Spindle mi-
         crotubules (arrows) first appear ra-
         diating from two adjacent spindle
         pole bodies (SPB). Later, such
         half spindles will interact to form
         the complete bipolar spindle.
         X 90,000. (From Schrével, <u>et</u> <u>al</u>.,
         1977; reproduced by permission of
         Academic Press, Inc.)

genetic implications of mitotic variations within the fungi,
our understanding of mitotic variations in the Protista will
be advanced. Either way, we win.

ACKNOWLEDGMENTS

     I am grateful to all those who provided their electron
micrographs for use in this paper and to M.S. Fuller and
J.L. Scott who shared their unpublished observations with me.
Helpful criticisms by I.B. Heath and R.B. Nicklas were very
much appreciated. This paper was written while I was supported by National Science Foundation Grant PCM 76-20131.

REFERENCES

Aikawa, M., Sterling, C.R., and Rabbege, J. (1972). Proc.
    Helminthol. Soc. Wash. 39, Spec. Issue, 174.
Ainsworth, C.G. (1971). "Dictionary of the Fungi", 6th Ed.
    Commonw. Mycol. Inst., Kew, Surrey, England.
Aist, J.R., and Williams, P.H. (1972). J. Cell Biol. 55,
    368.
Bajer, A.S., and Molè-Bajer, J. (1972). "Spindle Dynamics
    and Chromosome Movement", International review of Cyto-
    logy, Suppl. 3, 271 pp. Academic Press, New York.
Bartnicki-Garcia, S. (1970). In "Phytochemical Phylogeny"
    (J.B. Harborne, ed.), pp. 81-103. Academic Press,
    New York.
Beckett, A., and Crawford, R.M. (1970). J. Gen. Microbiol.
    63, 269.
Bessey, E.A. (1950). "Morphology and Taxonomy of Fungi",
    Blakiston Company, Philadelphia.
Bland, C.E., and Lunney, C.Z. (1975). Cytobiologie 11, 382.
Borisy, G.G., Peterson, J.B., Hyams, J.S., and Ris, H. (1975).
    J. Cell Biol. 67, 38a (abstract).
Bouck, G.B., and Brown, D.L. (1973). J. Cell Biol. 56, 340.
Brinkley, B.R., and Cartwright, J. (1971). J. Cell Biol.
    50, 416.
Brinkley, B.R., Stubblefield, E., and Hsu, T.C. (1967). J.
    Ultrastruct. Res. 19, 1.
Byers, B., and Goetsch, L. (1973). Cold Spring Harbor Symp.
    Quant. Biol. 38, 123.
Byers, B., and Goetsch, L. (1975). Proc. Nat. Acad. Sci.,
    U.S. 72, 5056.
Cachon, J., and Cachon, M. (1974). C.R. Acad. Sci., Ser. D
    278, 1735.
Cachon, J., and Cachon, M. (1977). Chromosoma 60, 237.
Chadefaud, M. (1975). Ann. Sci. Nat., Bot., 12e Serie, 16,
    217.
Chatton, E. (1937). "Titres et Travaux Scientifique",
    Sète, Sottano.
Chong, J., and Barr, D.J.S. (1974). Can. J. Botany 52, 1197.
Cleveland, L.R., Hall, S.R., Sanders, E.P., and Collier, J.
    (1937). Mem. Amer. Acad. Arts Sci. 17, 185.
Cohen, S.S. (1970). Amer. Sci. 58, 281.
Cohen, S.S. (1973). Amer. Sci. 61, 437.
Cuzin, F., and Jacob, F. (1967). In "Regulation of Nucleic
    Acid and Protein Biosynthesis" (V.V. Koningsberger and
    L. Bosch, eds.), pp. 39-50. Elsevier, Amsterdam.
Day, A.W. (1972). Can. J. Botany 50, 1337.
Demoulin, V. (1974). Bot. Rev. 40, 315.

Denison, W.C., and Carroll, G.C.   (1966). _Mycologia_ 48, 249.
Dietz, R.   (1969). _Naturwissenschaften_ 56, 237.
Dubremetz, J.-F.   (1973). _J. Ultrastruct. Res._ 42, 354.
Erwin, J.A.   (1973). "Lipids and Biomembranes of Eukaryotic
    Microorganisms", Academic Press, New York.
Febvre, J.   (1977). _J. Ultrastruct. Res._ 60, 279.
Flavell, R.   (1972). _Biochem. Genet._ 6, 275.
Forer, A.   (1974). _In_ "Cell Cycle Controls" (G.M. Padilla,
    I.L. Cameron, and A.M. Zimmerman, eds.), pp. 319-336.
    Academic Press, New York.
Franke, W.W., and Reau, P.   (1973). _Arch. Mikrobiol._ 90,
    121.
Fritsch, F.E.   (1935). "The Structure and Reproduction of
    the Algae", Vol. 1. Cambridge University Press, London
    and New York.
Fuller, M.S.   (1976). _Int. Rev. Cytol._ 45, 113.
Fuller, M.S., and Calhoun, S.A.   (1968). _Z. Zellforsch._
    Mikrosk. Anat. 87, 526.
Fuller, M.S. and Reichle, R.E.   (1965). _Mycologia_ 57, 946.
Girbardt, M.   (1968). _Symp. Soc. Exp. Biol._ 22, 249.
Girbardt, M.   (1971). _J. Cell Sci._ 9, 453.
Gordon, C.M.   (1977). _J. Cell Sci._ 24, 81.
Grassé, P.-P.   (1952). "Traité de Zoologie", Vol. I, Part 1.
    Masson, Paris.
Gull, K., and Newsam, R.J.   (1975). _Protoplasma_ 83, 247.
Gull, K., and Trinci, A.P.J.   (1974). _Trans. Brit. Mycol._
    Soc. 63, 457.
Harder, D.E.   (1976a). _Can. J. Botany_ 54, 981.
Harder, D.E.   (1976b). _Can. J. Botany_ 54, 995.
Harper, R.A.   (1905). Carnegie Institute of Washington Pub-
    lication No. 37.
Heath, I.B.   (1974a). _In_ "The Cell Nucleus" (H. Busch, ed.),
    pp. 487-515. Academic Press, New York.
Heath, I.B.   (1974b). _J. Cell Biol._ 60, 204.
Heath, I.B.   (1974c). _Mycologia_ 66, 354.
Heath, I.B.   (1975). _Biosystems_ 7, 351.
Heath, I.B., and Greenwood, A.D.   (1968). _J. Gen. Microbiol._
    53, 287.
Heath, I.B., and Greenwood, A.D.   (1970). _J. Gen. Microbiol._
    62, 139.
Heath, I.B., and Heath, M.C.   (1976). _J. Cell Biol._ 70, 592.
Hollande, A., and Carruette-Valentin, J.   (1970). _C.R. Acad._
    Sci., _Ser. D_ 270, 1476.
Hollande, A., and Carruette-Valentin, J.   (1971). _Protisto-_
    logica 7, 5.
Hollande, A., and Valentin, J.   (1968a). _C.R. Acad. Sci._,
    Ser. D 266, 367.
Hollande, A., and Valentin, J.   (1968b). _C.R. Acad. Sci._,

Ser. D 267, 1383.

Hollande, A., Cachon, J., and Cachon, M. (1969). C.R. Acad. Sci., Ser. D 269, 179.

Honigberg, B.M. (Chairman) (1964). J. Protozool. 11, 7.

Howard, K.L., and Moore, R.T. (1970). Botan. Gaz. 131, 311.

Ichida, A.A., and Fuller, M.S. (1968). Mycologia 60, 141.

Inoué, S., and Ritter, H. (1975). In "Molecules and Cell Movement" (S. Inoué and R. Stephens, eds.), pp. 3-30. Raven Press, New York.

Inoué, S., and Sato, H. (1967). J. Gen. Physiol. 50, Suppl., 259.

Kelley, G.L., and Hammond, D.M. (1972). Z. Parasitenk. 38, 271.

Klein, R.M., and Cronquist, A. (1967). Quart. Rev. Biol. 42, 105.

Kohlmeyer, J. (1975). Bioscience, 25, 86.

Kubai, D.F. (1973). J. Cell Sci. 13, 511.

Kubai, D.F. (1975). Int. Rev. Cytol. 43, 167.

Kubai, D.F., and Ris, H. (1969). J. Cell Biol. 40, 508.

Leadbeater, B., and Dodge, J.D. (1967). Arch. Mikrobiol. 57, 239.

Leedale, G.F. (1968). In "The Biology of Euglena" (D.E. Buetow, ed.), Vol. 1. pp. 185-242. Academic Press, New York.

Leedale, G.F. (1970). Ann. N. Y. Acad. Sci. 175, 429.

LéJohn, H.B. (1974). In "Evolutionary Biology" (T. Dobzhansky, M.K. Hecht, and W.C. Steere, eds.), Vol. 7. pp. 79-125. Plenum Press, New York.

Lerbs, V. (1971). Arch. Mikrobiol. 77, 308.

Lerbs, V., and Thielke, C. (1969). Arch. Mikrobiol. 68, 95.

Lessie, P.E., and Lovett, J.S. (1968). Amer. J. Botany 55, 220.

Liebowitz, P.J., and Schaechter, M. (1975). Int. Rev. Cytol. 41, 1.

Lu, B.C. (1967). J. Cell Sci. 2, 529.

McCully, E.K., and Robinow, C.F. (1971). J. Cell Sci. 9, 475.

McCully, E.K., and Robinow, C.F. (1972a). J. Cell Sci. 10, 857.

McCully, E.K., and Robinow, C.F. (1972b). J. Cell Sci. 11, 1.

McCully, E.K., and Robinow, C.F. (1973). Arch. Mikrobiol. 94, 133.

McDonald, K. (1972). J. Phycol. 8, 156.

McIntosh, J.R., and Landis, S. (1971). J. Cell Biol. 49, 468.

McIntosh, J.R., Cande, W.Z., and Snyder, J.A. (1975). In

"Molecules and Cell Movement" (S. Inoué and R. Stephens, eds.), pp. 31-76. Raven Press, New York.
McIntosh, J.R., Hepler, P.K., and Van Wie, D.G. (1969). Nature (London) 224, 659.
McNitt, R. (1973). Can. J. Botany 51, 2065.
Manton, I. (1964). J. Roy. Microsc. Soc. 83, 317.
Margulis, L. (1970). "Origin of Eukaryotic Cells", Yale University Press, New Haven.
Mayron, R., and Wise, D. (1976). Chromosoma 55, 69.
Mazia, D. (1961). In "The Cell" (J. Brachet and A.E. Mirsky, eds.), Vol. 3, pp. 77-412. Academic Press, New York.
Moens, P.B. (1971). Can. J. Microbiol. 17, 507.
Moens, P.B. (1976). J. Cell Biol. 68, 113.
Moens, P.B., and Church, K. (1977). Chromosoma 61, 41.
Moens, P.B., and Rapport, E. (1971). J. Cell Biol. 50, 344.
Moorman, G.W. (1976). Mycologia 68, 902.
Motta, J.J. (1969). Mycologia 61, 873.
Mueller, E. and von Arx, J.A. (1973). In "The Fungi" (G.C. Ainsworth, F.K. Sparrow and A.S. Sussman, eds.), Vol. 4A, pp. 87-132. Academic Press, New York.
Nicklas, R.B. (1971). Advan. Cell Biol. 2, 225-297.
Oakley, B.R., and Dodge, J.D. (1974). J. Cell Biol. 63, 322.
Oestergren, G. (1949). Hereditas (Lund) 35, 445.
Oestergren, G., Molè-Bajer, J., and Bajer, A. (1960). Ann. N. Y. Acad. Sci. 90, 381.
Olive, L.S. (1975). "The Mycetozoans", Academic Press, New York.
Olson, L.W. (1974). C.R. Trav. Lab. Carlsberg 40, 125.
Ott, D.W., and Brown, R.M. (1972). Br. phycol. J. 7, 361.
Peterson, J.B., and Ris, H. (1976). J. Cell Sci. 22, 219.
Peterson, J.B., Gray, R.H., and Ris, H. (1972). J. Cell Biol. 53, 837.
Peyrière, M. (1971). C.R. Acad. Sci., Ser. D 273, 2071.
Pickett-Heaps, J.D. (1969). Cytobios 1, 257.
Pickett-Heaps, J.D. (1971). Cytobios 3, 205.
Pickett-Heaps, J.D. (1972). Cytobios 5, 59.
Pickett-Heaps, J.D. (1974). Biosystems 6, 37.
Poon, N.H., and Day, A.W. (1976a). Can. J. Microbiol. 22, 495.
Poon, N.H., and Day, A.W. (1976b). Can. J. Microbiol. 22, 507.
Porter, K.R. (1966). Principles Biomol. Organ., Ciba Found. Symp., 1965. pp. 308-345.
Powell, M. (1975). Can. J. Botany 53, 627.
Raff, R.A., and Mahler, H.R. (1972). Science 177, 575.
Raju, N.B., and Lu, B.C. (1973). J. Cell Sci. 12, 131.
Rapport, E. (1971). Can. J. Genet. Cytol. 13, 55.

Raven, P.H.   (1970).   Science 169, 641.
Rickards, G.K.   (1975).   Chromosoma 49, 407.
Ris, H., and Kubai, D.F.   (1974).   J. Cell Biol. 60, 702.
Robinow, C.F., and Caten, C.E.   (1969).   J. Cell Sci. 5, 403.
Robinow, C.F., and Marak, J.   (1966).   J. Cell Biol. 29, 129.
Rooney, L., and Moens, P.B.   (1973).   Can. J. Microbiol. 19, 1383.
Roos, U.-P.   (1975).   J. Cell Biol. 64, 480.
Rosenberger, R.F., and Kessel, M.   (1968).   J. Bacteriol. 96, 1208.
Ryter, A.   (1968).   Bacteriol. Rev. 32, 39.
Sabatini, D.D., Bensch, K., and Barrnett, R.J.   (1963).   J. Cell Biol. 17, 19.
Sagan, L.   (1967).   J. Theor. Biol. 14, 225.
Savile, D.B.O.   (1968).   In "The Fungi" (G.C. Ainsworth, and A.S. Sussman, eds.), Vol. 3, pp. 649–675.   Academic Press, New York.
Schrader, F.   (1953).   "Mitosis" 2nd Ed.   Columbia University Press, New York.
Schrantz, J.-P.   (1967).   C.R. Acad. Sci., Ser. D 264, 1274.
Schrantz, J.-P.   (1970).   Rev. Cytol. Biol. Veg. 33, 1.
Schrével, J., Asfaux-Foucher, G., and Bafort, J.M.   (1977).   J. Ultrastruct. Res. 59, 332.
Scott, J.L., and Thomas, J.P.   (1975).   J. Phycol. 11, 474.
Segal, E., and Eylan, E.   (1975).   Microbios 12, 111.
Setliff, E.C., Hoch, H.C., and Patton, R.F.   (1974).   Can. J. Botany 52, 2323.
Siebert, A.E., and West, J.A.   (1974).   Protoplasma 81, 17.
Slankis, T., and Gibbs, S.P.   (1972).   J. Phycol. 8, 243.
Southall, M.A., Motta, J.J., and Patterson, G.W.   (1977).   Am. J. Botany 64, 246.
Soyer, M.O.   (1969).   C.R. Acad. Sci., Ser. D 268, 2082.
Soyer, M.O.   (1971).   Chromosoma 33, 70.
Sparrow, F.K.   (1960).   "Aquatic Phycomycetes" 2nd ed., University of Michigan Press, Ann Arbor.
Stanier, R.Y.   (1970).   Symp. Soc. Gen. Microbiol. 20, 1–38.
Stanier, R.Y., and van Niel, C.B.   (1962).   Arch. Mikrobiol. 42, 17.
Sun, N.C., and Bowen, C.C.   (1972).   Caryologia 25, 471.
Syrop, M., and Beckett, A.   (1976).   Mycologia 68, 902.
Tanaka, K.   (1970).   Protoplasma 70, 423.
Tiffney, B.H., and Barghoorn, E.S.   (1974).   Occasional Papers of the Farlow Herbarium 7, 1.
Uzzell, T., and Spolsky, C.   (1974).   Amer. Sci. 62, 334.
Van Winkle, W.B., Biesele, J.J., and Wagner, R.P.   (1971).   Can. J. Genet. Cytol. 13, 873.
Wells, K.   (1970).   Mycologia 62, 761.
Whisler, H.C., and Travland, L.B.   (1973).   Arch. Protistenk.

   115, 69.
Whittaker, R.H.   (1969).   Science 163, 150.
Yamanouchi, S.   (1906).   Botan. Gaz. 42, 401.
Zickler, D.   (1970).   Chromosoma 30, 287.

GENUS INDEX

A

Achlya, 93, 138
Aessosporon, 97, 143, 211
Albugo, 16
Allomyces, 5, 6, 94, 142, 161, 207
Amphidinium, 181
Ancylistes, 95, 105, 127, 209, 210
Aphanomyces, 93, 99
Apodachlya, 94
Arbacia, 26
Arcyria, 93, 137
Armillaria, 97, 115
Ascobolus, 95, 101, 102, 107, 137, 143, 211
Aspergillus, 95, 102, 138, 141, 144, 145, 147, 151–153, 199, 202
Asterias, 68, 71

B

Basidiobolus, 8, 95, 106, 127, 142, 151, 159, 160, 209, 210
Blastocladiella, 94, 207
Blastodinium, 181
Boletus, 96, 102
Botrytis, 152

C

Catenaria, 94, 115, 207, 215
Cavostelium, 92
Ceratiomyxa, 93
Ceratocystis, 95, 143
Chaetomium, 102
Chara, 8
Clastoderma, 93
Cochliobolus, 96, 117
Coleosporium, 12
Collozoom, 219
Conidiobolus, 95, 127, 128, 175, 194
Coprinus, 5, 97, 104, 114, 137
Corticium, 12
Cryptothecodinium, 181–183

D

Dictyostelium, 92, 105, 115, 121, 123, 125, 126, 129, 134, 218
Didymium, 93, 113, 142
Dipodascus, 95

E

Echinostelium, 93, 137
Echinus, 25
Eimeria, 222
Endogone, 142
Entophlyctis, 94, 116, 140, 194, 207
Erysiphe, 96, 101

F

Fomes, 158
Fusarium, 5, 96, 131, 143, 145, 158, 188, 192, 199

G

Griffithsia, 217
Gymnosporangium, 97

H

Haemanthus, 43, 45–47, 49, 61–64
Haplozoon, 181
Harpella, 105
Harpochytrium, 94, 194, 207

L

Labyrinthula, 91, 92, 98
Leucosporidium, 96, 143, 211
Lipomyces, 95

SUBJECT INDEX

A

Acantharia, 219, 220, 221
Acrasiales, general, 92–98, 105, 118
Actin, 21, 34–80, 115, 140, 141, 145, 156
Algae, general, 8, 35, 106, 214–218
Amiprophos-methyl, 154
Amoeboid movement, 35, 41
Anaphase, 22, 23, 43, 49, 67, 133–137, 139, 145, 188–190, 204
Antibodies
  as inhibitors, 52–59
  as "stains", 65–80
Archontosome, 90
Ascomycetes, general, 92–98, 100–102, 104, 106, 126, 137, 152, 188, 190, 194, 195, 206, 208–213, 219
ATPases, 52, 54, 56–58, 141

B

Basidiomycetes, general, 92–98, 102–104, 114, 117, 126, 135, 137, 143, 152, 176, 188, 206, 211–213
Benomyl see M.B.C.
Birefringence, 23–28, 34
Blood platelets, 40

C

Calcium, 140
Camphor, 148
Cell cleavage, 35
Centrioles
  general, 10, 90–99, 107, 109, 119, 124, 125, 130, 131, 149, 198, 206, 207, 215
  migration, 91–98
  replication, 91–98
Centromere see Kinetochore
Centrosomes, 1
Chick cells, 40
Chromatin, 2–11, 105, 126, 127

Chromosomes, 6–8, 14, 17, 22, 23, 43, 46, 92–98, 121, 122, 126, 128, 131–133, 137–139, 141, 156, 179, 180, 182–185, 189, 190, 193–195, 197, 198, 200, 201, 203–205
Chytridiomycetes, 92–98, 116, 117, 139, 195, 201, 206–209, 214, 215
Colcemid see Colchicine
Colchicine, 141, 147–150
Crane flies, 42–44, 46, 49–51, 75
Cyclic A.M.P., 160–162

D

Deuteromycetes, general, 126
Dinoflagellates, 9, 180–184, 214, 216
DNA, 6, 7, 91, 107, 108, 137, 138, 152, 178, 202

E

Endoplasmic reticulum, 11, 16, 113, 116

F

Fixation, 2, 4, 8, 16, 28, 29, 39, 42, 50, 61, 62, 65, 66, 68, 101, 114, 192
Fossil record, 177

G

Griseofulvin, 151

H

Heavy meromyosin, 36–50, 59, 60, 72, 75–80
Histones, 138, 161
Hydrostatic pressure, 155
Hyphochytridiomycetes, general, 92–98, 116, 117, 139, 206, 208, 209, 214

I

Interphase, 2, 10–12, 14, 91
I.P.C., 155